Multi-Objective Management of Saltwater Intrusion in Groundwater: Optimization under Uncertainty

Multi-Objective Management of Saltwater Intrusion in Groundwater: Optimization under Uncertainty

PROEFSCHRIFT

ter verkrijging van de graad van doctor
aan de Technische Universiteit Delft,
op gezag van de Rector Magnificus prof.dr.ir. J.T. Fokkema,
voorzitter van het College voor Promoties,
in het openbaar te verdedigen op dinsdag 27 april 2004 om 10.30 uur

door

Thuan Minh TRAN

Master of Science - Wageningen Agricultural University
geboren te Can Tho, Viet Nam

Dit proefschrift is goedgekeurd door de promotor:
Prof.dr.ir. C. van den Akker

Samenstelling promotiecommissie:

Rector Magnificus, voorzitter
Prof.dr.ir. C. van den Akker, Technische Universiteit Delft, promotor
Prof.ir. E. van Beek, Technische Universiteit Delft
Prof.dr.ir. A.W. Heemink, Technische Universiteit Delft
Prof.dr.ir. C. Roos, Technische Universiteit Delft
Prof.dr.ir. A. Leijnse, Wageningen Universiteit
Prof.dr.ir. A.M. Elfeki, Mansoura University, Egypt
Prof.dr.ir. M.Q. Le, Can Tho University, Can Tho, Viet Nam

Dr. E.J.M. Veling en Dr. R.J.Schotting hebben als begeleiders in belangrijke mate aan de totstandkoming van het proefschrift bijgedragen.
Het onderzoek in dit proefschrift is een onderdeel van het project "Development of Curricula in Civil and Mechanical Engineering", Can Tho University, Can Tho, Viet Nam en werd financieel mogelijk gemaakt door het MHO-programma onder beheer bij Nuffic en gecoördineerd door CICAT, Delft.

Published and distributed by: DUP Science
DUP Science is an imprint of
Delft University Press
P.O.Box 98
2600 MG Delft
The Netherlands
Telephone: +31 15 2785678
Telefax: +31 15 2785706
E-mail: info@library.tudelft.nl

ISBN: 90-407-2480-6

Printed in The Netherlands.

Acknowledgment

Albeit performing this thesis is an individual task, however, I would like to thank very much a number of people who have given great helps to me for the success of this work.

First of all I would like to thank prof.ir. R. Brouwer for receiving me early on in 1995 and secondly prof.dr.ir. C. van den Akker for accepting to be my promoter in 1997 and for introducing me to the section of Hydrology and Ecology. During the years I carried out this thesis they both provided me with the facilities of the department.

Special thanks are due to dr. E.J.M. Veling and dr. R.J. Schotting (my supervisors) who initially proposed the valuable idea for my topic of research and encouraged me to pursue this research even when I encountered difficulties in my job and life. Especially for dr. E.J.M. Veling without your help my thesis could not have been completed.

Sincere gratitude is expressed to prof.dr. T. Terlaky, prof.dr.ir. C. Roos, dr. E. de Klerk for discussing and giving valuable hints to solve the problems in the quadratic cone optimization. I would like to take this opportunity for memorizing dr. J.F. Sturm who has developed the SeDuMi package which I have applied and which was very instrumental in my research. With him I had a lot of contact by e-mail for discussing problems during the last two years of my work.

My sincere appreciation is forwarded to prof.dr.ir. C. van den Akker, prof.dr.ir. C. Roos, prof.dr.ir. A.W. Heemink and prof.dr.ir. A. Leijnse whose constructive criticism made it possible to produce this thesis.

I am indebted to my rector of Can Tho University, Viet Nam, prof.dr.ir. Le Quang Minh who allowed and created a good opportunity for my leaving Can Tho University to do my Ph.D. for several years. Thanks are also forwarded to my colleague, Mr. Nguyen Van Tinh for shouldering my leadership of the Civil Engineering Department in Can Tho University.

My thanks are also to my former roommates Twan Gielen, Willem Jan Zaadnoordijk for their helpful discussion and warm friendship and to Neeltje van de Wiel, for introducing me to software, books and being kind and friendly. I would like to thank dr. R.R.P. van Nooyen for his kindness to install software on my computer. My sincere thanks are sent to Margreet Evertman and Hanneke de Jong, the secretaries of the section of Hydrology and Ecology, for your administrative responsibilities.

My gratefulness is to Nuffic – MHO for making my study possible through financial assistance and to CICAT staff members, drs. P. Althuis, dr. J.A. van Dijk, Veronique van der Varst, Rob Nievaart, Franca Post, Theda Olsder, Kate Landzaat and Christel Crone for their logistic arrangement and warm friendship during my stay in Delft.

Last but not least, I would like to thank my wife Phuong for her tolerance, encouragement and well management of the family while I was away from home. My son Quan, I was very delighted that you were grown up and more self-confident every time I saw you again. And I hope my very young daughter Nhu will remember and recognize her daddy when I come back home this time.

Dedicated to my parents, my wife Phuong and my children Quan and Nhu.

Contents

Part I

Main Thesis

Chapter 1

Introduction

1.1 General

Groundwater aquifers are an important resource in coastal regions. In many areas where there is no fresh surface water from rivers or reservoirs, the development of ground water resources is practically the only alternative to storage of rainwater.

However, the coastal aquifers are very vulnerable to the seawater intrusion through the overdraft of groundwater exploitation or insufficient recharge from upstream, etc. In order to control sea water intrusion, varying forms of saltwater intrusion management models have been studied. They address the optimal groundwater pumping and recharge schedules with or without surface water supplies for conjunctive use.

Problems of salt-intrusion of groundwater have become a considerable concern in many countries which have coastal areas. There has been much research relating to saltwater intrusion in regions such as: in the Mediterranean coast of Israel by Shamir, Bear and Gamliel (1984), in the Waialae aquifer of southern Oahu, Hawaii by Essaid (1986), Emch and Yeh (1998), in southern Oahu, Hawaii by Souza and Voss (1987), in Hallandale, Florida by Andersen et al. (1988), in the Yun Lin Basin, Taiwan by Willis and Finney (1988), in the Soquel-Aptos basin, Santa Cruz County, California by Essaid (1990a, 1990b), in the Jakarta Basin by Finney et al. (1992), in the Dutch coast by Oude Essink (1996), etc. Amongst these, the aquifer systems are characterized by either single layer (unconfined) or multiple layers with varying hydraulic properties.

Two general approaches have been used to analyze saltwater intrusion in coastal aquifers: the disperse interface and sharp interface approaches. The disperse interface approach explicitly represents a transition zone that is a mixing zone (brackish water) of the freshwater and salt water within an aquifer due to the effects of hydrodynamic dispersion. In the transition zone there is a gradual change in density. Models that incorporate simulation of the transition zone may require simultaneous solution of the governing fluid flow and solute transport equations. The second approach is based on the simplification of the thin transition zone relative to the dimension of the aquifer. The freshwater and saltwater are considered to be two immiscible fluids of different constant densities. The studies based on this approach are modeled by only solving the groundwater flow equation (see Emch and Yeh, 1998).

In fact, reclamation of saltwater-polluted aquifers is a groundwater quality management model, which involves implementing remedial control measures for the rehabilitation of contaminated groundwater supplies. These options include physical containment, in situ rehabilitation, and withdrawal followed by treatment and use (Lehr and Nielsen, 1982). Physical containment systems prevent the flow of contaminated groundwater by controlling the flow field via the slurry trenches, cut-off walls, or grout curtains, or by altering the circulation pattern of the aquifer system through pumping or injection. Typically, aquifer rehabilitation involves an injection and recharge system. Withdrawal and treatment does not, however, exploit or utilize the aquifer's assimilative waste capacity; it simply removes the contaminated water from the groundwater system (see Willis and Yeh, 1987).

For this type of problem the mathematical optimization model must incorporate the groundwater flow (see Finney et al., 1992) and might also be linked to with the contamination transport simulation (see Gorelick et al., 1984).

1.2 The groundwater use in the Mekong Delta of Viet Nam

The Mekong Delta is an important region in Viet Nam; it has an area of 3.9 million ha of which 2.4 million ha is currently used for agriculture. Due to such a very large area of arable land, the Mekong Delta is one of the major rice production areas of Viet Nam. Economic reform leading to a market orientation economy in Viet Nam has brought positive progress. Food production has been improved and the agrarian conditions have been changed. According to the Master Plan, the potential for expansion of agricultural land is approximately 0.2 million ha or an increase of some 8% of the presently cultivated area. The Delta development needs to increase primary sector production. This cannot be achieved simply by reclamation of new land. Among other methods, intensification of land use has to be a key factor in achieving the required growth. Especially single rice cropping can be extended to double (rice) cropping and there is ample scope for crop diversification. Double rice cropping will lead to higher water requirements although this may be counteracted to some extent by the crop diversification; upland crops and fruit trees require less water per unit of land. The Master Plan for the Development of the Mekong Delta has identified the following trends affecting agriculture and the environment:

- intensification of land used for cultivation to meet increased demands for production.
- crop diversification.
- dependence on fresh water, in which river flows are becoming more scarce due to upstream allocation.
- vulnerability to ecological impoverishment.

The inventory of domestic water supply carried out by the Master Plan leads to the following conclusion:

- The present situation in the rural areas is far from ideal; some 8 million people living in parts of the Delta where the surface water is either saline or acid have to obtain their drinking water from a distance of more than 10 km in the dry season.
- In the urban areas, surface water is the traditional source for drinking water. Due to the lack of reagents, poor maintenance and the high sediment content of the river water in the flood season, the quality of the treated water is often poor. The bacteriological quality of the river water, in particular in the densely populated areas, is also poor.
- In general, groundwater is an attractive alternative because it has good bacteriological quality.
- In several towns in the coastal area surface water is transported over long distances (20 km and more). However, the salinity and turbidity of the water increase considerably during transport. For these towns, water supply from groundwater would mean a considerable improvement.

For the rural areas, development of ground water resources is practically the only alternative to storage of rainwater and it is expected that the number of small wells will increase drastically in the near future (some 19,000 small wells had been drilled for water supply by 1990).

However, the available geo-hydrological data is generally insufficient to determine the exact, local effects of groundwater abstraction. It is recommended to exploit ground water resources for small scale abstraction.

The subsoil of the Delta contains huge quantities of groundwater. Nevertheless, its exploitation is constrained by three factors:

a. the quality of the five aquifers, mainly by high salt concentration,

b. the permeability of the aquifers, and

c. the fresh water recharge of the aquifers, which determines the safe yield. (Anonymous, 1993 (Master Plan of Mekong Delta)).

As described by Michael (1971), an upper section of recent alluvium and a lower section of older alluvium underlie the Mekong Delta. The older alluvium contains a permeable artesian zone called the 100-metre aquifer or upper Pleistocene aquifer, which is the most productive groundwater reservoir in Viet Nam. Tested well capacities range from about 32.993 m^3/h - 144.399 m^3/h; more efficiently designed wells should produce in the range of 114 m^3/h - 227 m^3/h from this aquifer. Part of the 100-metre aquifer is intruded by seawater. The most feasible plan for development of the Mekong Delta may involve the conjunctive use of surface water and groundwater of 100-metre aquifer, even though induced recharge and a groundwater barrier against seawater intrusion may be necessary.

For the surface water system, most of the water flows are uncontrolled, especially during the period August-October. The northern half of the Delta becomes inundated by floodwaters of Mekong and Bassac river; these waters fail to drain away, and ultimately become stagnant, a condition which helps lead to acidification of the soil to the point where the land becomes non-arable. When stream levels are lowest, generally during the period November-April, high tides force seawater far inland.

From earliest times, inhabitants of the Mekong Delta have relied upon surface water, groundwater from shallow wells, and stored rainwater to meet domestic demands. Agricultural demands are supplied almost entirely by surface water. Acute shortages are experienced locally during the dry season, when rainwater stores are depleted and shallow wells become saline or high in aluminum sulfate. Drilled wells have been introduced to supplement municipal supplies.

In the Mekong Delta, the ground water is used mainly for domestic water supply in wide areas that are either far from the Mekong river system or near the coastline where there is no fresh surface water supply. The hand pumped or small engine pumped wells are predominantly used, therefore their use is only on a small scale. The present groundwater abstraction for domestic and industrial use amounts to roughly 75,000 m^3/day for the urban centres and 90,000 m^3/day for the rural areas. The total groundwater abstraction in the Delta thus amounts to 165,000 m^3/day.

1.3 The study area

In the coastal areas, the regions that are located along the Mekong river mounts consisting of Tra Vinh, Ben Tre, Tien Giang provinces are generally selected for a study area. This is because it has been the major area where seawater intrudes farther inland through the river mounts in the dry season. Especially in the Ben Tre province, its aquifers have been intruded by saltwater far inland affecting the groundwater quality for drinking. For this particular area the geo-hydrological data are very scarce, and therefore the geo-hydrological properties of the aquifers are only given under the average values, as in Table 1.1.

Table 1.1: Geo-hydrological properties of the aquifers.

Aquifers	Specific yield [l/s.m]	Thickness [m]	Transmissibility [m^2/d]
Holocene (Q_{IV})	-	20 - 50	Not important
Pleistocene (Q_p)	0.41 - 3.942	133 - 164	800 - 1300
Pliocene (N_2)	0.1 - 1.5	≥120	300 - 440
Miocene (N_1)	0.2 - 0.9	≥100	550

Note: These properties of the aquifers are averaged for the whole area.

Topographically, the area is relatively flat and low; the average elevation of ground surface ranges from +0.5m to +3m. This might be endangered by the subsidence when the groundwater abstraction is more than the water recharge for the aquifers.

Since the whole area is subject to the monsoon weather, the two main seasons are formed in one year: the wet and the dry season. Surrounding the area are the rivers, My Tho and Ham Luong, which have big discharges in the wet season and small discharges in the dry season. Consequently, in the wet season the rivers play roles in evacuating the floodwater from upstream to the east sea and also supplying a considered amount of freshwater for the drinking demand of the area. On the other hand, in the dry season seawater intrudes through the estuaries into the rivers, resulting in the impossibility of the surface water supply system. Therefore, the groundwater supply, which is the only alternative to the surface water, can possibly be performed under the threat of saltwater intrusion in the dry period.

However, the groundwater exploitation in the area has not been properly managed. The artificial recharge for compensating the aquifers for the extracted water is not considered as important as it should be in order to prevent the further intrusion of saltwater in those aquifers. Nowadays, especially with the increase of groundwater abstraction, the requirement for artificial recharge is more than ever.

The options of feasible plans for salt-polluted aquifer reclamation in this area can be briefly drawn as follows:

- Physical containment systems prevent the saline groundwater flow by controlling the flow field by injection of the fresh surface water. The underground structures might be feasible for the superficial aquifers only, whereas the pumping or injection alternatives can be applied to many types of aquifers.

- In situ rehabilitation involves the fresh water injection for in-ground dilution and an artificial recharge system. It implies that the salt concentration in the salt-polluted aquifers will go down to a desired level given sufficient time and space. The salt-fresh interface or transition zone in its aquifer is also expected to move seaward corresponding to the reclamation progress.

- Withdrawal of the saltwater out of the aquifers may be advisable for a fast-progress reclamation of seawater-intrusion aquifers only when it can be assured that they will be supplemented by the recharge system of fresh water.

Among the alternatives, the artificial recharge through injection can be solved considerably if one can make use of the flood water in the Mekong rivers e.g. My Tho river for injecting into the aquifers in order to dilute saltwater or push back the salt/fresh water front seaward. During the peak flow periods which are from August to October the peak discharge of the Tien river[1], a Mekong river branch, is about 30,000 m^3/s and in the dry season in April its lowest flow is about 1,000 m^3/s. This also helps partly prevent the inundation of the upstream areas by evacuating a part of the surplus surface water flow through the injection during the flood season.

According to the data available at this moment the development of groundwater use takes place only in a limited area of about one third of the area. However some problems have already been raised as follows:

- How big will the safe yield be for allocating the pumping rates of the groundwater wells in each sub-region of the province.
- How to maintain the present quality for the fresh groundwater aquifers under the increase of the pumping demands of drilled wells.
- How to extend the fresh groundwater zone area for the heavily populated coastal regions where the water resources, both surface and groundwater, are polluted by seawater, especially in the dry season with high-salt concentration in drinking water.

1.4 Objectives

For developing the area, the alternative of using groundwater as a supplementary source in conjunctive use with surface water is a good possibility at present and in future planning. However, groundwater sources have posed some difficult problems to solve, such as the possible requirement for a groundwater barrier against seawater intrusion, and an induced recharge system in the upper area which would have to be fed by water treated for conformity with the environment of the 100-meter aquifer. It is also important to determine the degree to which subsidence might occur, and the means for its control in the event of widespread groundwater development, because of the relatively low elevations in the area (see Michael, 1971).

The management issues characterizing the conjunctive use problem are to determine roughly as follows:

a. The optimal pumping schedules (well locations and pumping rates) to satisfy given water demand.

b. The optimal injection schedules. This involves specifying the well locations and injection rates necessary to satisfy a flood evacuation demand and the desired level of the saltwater intrusion control.

c. The maximum waste input concentration (mainly sedimentation load, pH levels, chloride contents, etc.) in the injected water should satisfy the criteria in order to avoid well clogging. This issue will not be treated in this work.

[1] The discharge of the Tien River is roughly 50% of the total discharge of the Bassac and Mekong river system at Kratie.

1.5. The outline of the thesis

To achieve these objectives a number of studies will be carried out in this thesis. They are mainly:

Literature review

The available studies in the literature will be mentioned in Chapter 2. It helps the reader to follow as much as possible the progress of the related works that the thesis is probably based on.

Characterization of the responses of saltwater intrusion lengths with respect to stresses and transmissivities

This step is essential in model applications for management through the understanding of the salt/fresh water interface movement. Generally, through this sensitivity study, the non-linear response of the saltwater intrusion length with respect to stresses (extraction, injection rates) is verified and the distinct changes of the salt/fresh sharp interface with respect to the hydraulic conductivity variation are determined. This study will be presented in Chapter 3.

Introduction of the application of the second-order cone optimization (SOCO) programming technique and SeDuMi (an add-on for MATLAB) into the optimal management of saltwater intrusion in groundwater

The second-order cone optimization (SOCO) (or quadratic cone) programming technique together with the interior point method, which is the promising tool for solving large-scale optimization problems, will be recalled in Chapter 4. This methodology with the help of SeDuMi (an add-on for MATLAB, which is an optimization program package developed for linear, SOCO and semi definite programming) can be conveniently developed for the saltwater intrusion management problems, especially in cases where the coefficients of the objectives and constraints are in the uncertainty fields.

Development of a multi-objective management of saltwater intrusion in groundwater with deterministic and stochastic approaches as an add-on program for MATLAB

With the management of an aquifer system in coastal areas where the salt/fresh interface appears near the capture zone, it is necessary to include the objective for minimizing the saltwater intrusion length during the operation of the extraction and injection wells. Besides that the other objective for minimizing the operational costs is also included. The management problem is built by creating the linkages between the SHARP simulation model and SeDuMi optimizer through the response matrices under the MATLAB environment. This methodology will be mainly developed in Chapter 5 and applied in Chapters 6 and 7.

Application of the programs to the hypothetical and real world problems for the saltwater intrusion management

The multi-objective management of saltwater intrusion in groundwater programs is firstly applied to a hypothetical case in which the geometry of the modelled area is assumed to be symmetric. Either the mean value of the hydraulic conductivity is given (the deterministic case) or the random values for a number of realizations of the hydraulic conductivity are generated (the uncertainty case) for the input file of the simulation model. The candidate well locations, being the decision variables, are arranged in a symmetric way so that the results of the hypothetical problem in the deterministic case will help to check the validity of the programs when the optimal solution obtained is symmetric. This work will be done in Chapter 6.

The real world problem is addressed in one particular study area, selected from the coastal areas in the Mekong Delta in Viet Nam. The area is intruded by saltwater with the current interface position located near the pumping wells. For this particular area the available data are very scarce and only the averaged values of all aquifer properties are given. This management scheme for the prevention of saltwater intrusion by artificial injection, as proposed here, is new and can be applied in areas where there is a potential risk of saltwater intrusion. Therefore, the saltwater intrusion management problem presented in this work will give more insight in the sense of planning rather than the current management scheme of the study area.

Since the study area is subject to the tropical monsoon weather, there are two distinct seasons, the wet and the dry seasons. Under these circumstances a scheme of management is proposed – in this sense a so-called seasonal planning for the saltwater intrusion management problem. This consists of two managerial stages during one year– the first stage for the saltwater intrusion management during the wet season and the second stage during the dry season. The programs will run for the first stage and its solution will be attained. The salt/fresh interface that is simulated using the model with the optimal solution obtained in the first stage will be the initial interface for the second stage of the management problem. The real world problem will be carried out in Chapter 7.

Chapter 2

Literature Review

Saltwater intrusion into aquifers and ground water quality degradation by salinization are two of the most serious threats to fresh groundwater resources, which constitute an essential supply for human needs. This is especially true in the coastal areas and in dry climates (Custodio and Galofre, 1992). There were many studies of groundwater flow models to help the understanding and prediction of the behaviour of fresh and saline groundwater under a certain type of exploitation. These studies have been very important to the management of groundwater exploitation for over a century.

Salt water intrusion problems have been solved by using different methods, ranging from the basic Badon Ghyben-Herzberg principle with the sharp interface models to the more sophisticated theories with the solute transport models which take into account variable densities. The groundwater flow model is always a part of any model concerned with the movement of salt-fresh water interface and/or solute transport, whereas the solute transport model is necessary for solving most of the groundwater quality problems.

Emch and Yeh (1998) summarized the two general approaches that had been used to analyse saltwater intrusion in coastal aquifers. These are referred to as the sharp and disperse interface approaches. The first approach to the analysis of the saltwater intrusion problem is based on the simplifying assumption that the transition zone can be represented by a sharp interface. The fresh water and salt water are considered to be two immiscible fluids of different constant densities and the system is modelled using only the flow equation. The sharp interface assumption is considered reasonable when the width of the transition zone is small relative to the thickness of the aquifer. It is generally applicable to regional-scale systems. Sharp interface approaches have used one of two methods. The one-fluid method models freshwater dynamics only. It is assumed that the water table and the interface maintain continuous equilibrium and that the salt water is static. Alternatively, the two-fluid method may be used, in which coupled freshwater and salt water flow equations are solved simultaneously. The second approach is that the fresh water and saltwater zones within an aquifer are separated by a transition zone in which there is a gradual change in density. The disperse interface approach explicitly represents the presence of this zone. Models that incorporate simulation of the transition zone may require a simultaneous solution of the governing fluid flow and mass transport equations.

Management of saltwater intrusion problems often requires the use of non-linear optimization models due to the complexity of the governing equations. Solution techniques can be classified as either unconstrained, linearly constrained, or non-linearly constrained optimization methods (Emch and Yeh, 1998).

2.1 The groundwater flow models

2.1.1 The governing equation for the fresh groundwater flow in the one-fluid method

The governing groundwater flow equation below is restricted to fluids with a constant density or in cases where the differences in density or viscosity are extremely small or absent (Barends and Uffink, 1997). This equation is derived by mathematically combining a water balance equation with Darcy's law (Anderson and Woessner, 1992):

$$\frac{\partial}{\partial x}\left(K_x\frac{\partial h}{\partial x}\right)+\frac{\partial}{\partial y}\left(K_y\frac{\partial h}{\partial y}\right)+\frac{\partial}{\partial z}\left(K_z\frac{\partial h}{\partial z}\right)=S_s\frac{\partial h}{\partial t}-W^*, \qquad (2.1.a)$$

where:

K_x, K_y, K_z are components of the hydraulic conductivity tensor [LT^{-1}],

S_s is the specific storage [L^{-1}],

W^* is the general sink/source term that is intrinsically positive and defines the volume of inflow to the system per unit volume of aquifer per unit of time [T^{-1}],

h is the groundwater head [L],

x, y, z are the Cartesian coordinates [L],

t is time [T].

The solution of the above equation are the fresh groundwater heads with which the location of salt/fresh water interface will be calculated by the basic Badon Ghyben-Herzberg principle:

$$h_s=\left[\gamma_f/\left(\gamma_s-\gamma_f\right)\right]h_f\equiv\delta h_f \qquad (2.1.b)$$

where:

γ_f, γ_s specific weight of fresh and salt water [$M\,L^{-2}\,T^{-2}$]

h_f, h_s fresh water head above sea level, and static salt water head at interface [L]

2.1.2 The governing equation for the groundwater flow describing multiphase flow in the two-fluid method

Cases where density differences play a role and may not be neglected are encountered, for example, in coastal aquifers (salt/fresh water) (Barends and Uffink, 1997). Here the governing groundwater flow equation will have the form of the density dependent groundwater flow, which takes into account the variable density. The study and interpretation of variable density groundwater flow has attracted the attention of many researchers and engineers for a long time (Custodio, 1992).

The flow of fluids of different densities may involve miscible fluids, which mix and combine readily, or immiscible fluids, which do not. The governing equations describing two-fluid flow which is considered as the movement of an immiscible fluid can be written according to Bear (1979) and Essaid (1986):

$$\left[S_f b_f+\phi(a+\delta)\right]\frac{\partial h_f}{\partial t}-\phi(1+\delta)\frac{\partial h_s}{\partial t}=\frac{\partial}{\partial x}\left(b_f K_{fx}\frac{\partial h_f}{\partial x}\right)-\frac{\partial}{\partial y}\left(b_f K_{fy}\frac{\partial h_f}{\partial y}\right)+Q_f+R_f, \qquad (2.2.a)$$

and

$$\left[S_s b_s + \phi(1+\delta)\right]\frac{\partial h_s}{\partial t} - \phi\delta\frac{\partial h_s}{\partial t} = \frac{\partial}{\partial x}\left(b_s K_{sx}\frac{\partial h_s}{\partial x}\right) - \frac{\partial}{\partial y}\left(b_s K_{sy}\frac{\partial h_s}{\partial y}\right) + Q_s + R_s, \qquad (2.2.b)$$

where the interface elevation can be calculated from the fresh and saltwater heads:

$$z_i = (1+\delta)h_s - \delta h_f, \qquad (2.2.c)$$

where

h_f, h_s	are the freshwater head and saltwater hydraulic head [L],
S_f, S_s	freshwater and saltwater specific storage [1/L],
b_f, b_s	average saturation thickness of freshwater and saltwater zones [L],
ϕ	the effective porosity,
δ	$= \gamma_f / (\gamma_s - \gamma_f)$,
γ_f, γ_s	freshwater and saltwater specific weight [M L^{-2} T^{-2}],
R_f, R_s	fresh and saltwater leakage across top and bottom of aquifer [L/T],
Q_f, Q_s	fresh and saltwater source/sink terms (pumping, recharge) rates [L/T],
K_{fx}, K_{fy}	freshwater hydraulic conductivities [L/T],
K_{sx}, K_{sy}	saltwater hydraulic conductivities [L/T],
a	$a = 1$ if the aquifer is unconfined; $a = 0$ if the aquifer is confined.

For solving the dynamic flow of saltwater intrusion, the coupled freshwater and saltwater flow equations are solved simultaneously.

2.1.3 Simulation models

In order to obtain the groundwater head solution, the simulation models, which are based on the mathematical models with certain simplifying assumptions for the flow domain and its boundaries, will be solved by either analytical or numerical methods. At present, a large number of mathematical models are available, which are capable of handling fresh and saline groundwater flow in aquifer systems. They are subdivided into *analytical* and *numerical models* (see also Oude Essink, 1996).

When simplified, the groundwater flow equation (2.1a) might be solved analytically. The simplifications usually involve assumption of homogeneity and one- or two-dimensional flow. Except for applications to well hydraulics, analytical solutions for flow problems are not widely used in practical application. Numerical solutions are much more versatile and with the widespread availability of computers, are now easier to use than some of the more complex analytical solutions (Anderson and Woessner, 1992).

According to Oude Essink (1996), the numerical methods which are in combination with the sharp interface models for solving the salt-fresh groundwater flow are: *finite differences, finite elements, the boundary integral equation method, analytical elements and the method based on the vortex theory which has an analytic character.*

The sharp interface models

These models are based on the Badon Ghyben-Herzberg principle that assumes a sharp interface between fresh and saline groundwater, which is able to represent the actual situation.

The one-fluid models

The models are based on freshwater dynamics only. These were used by Glover (1959), Henry (1959), Shamir and Dagan (1971), Volker and Rushton (1982), Ayers and Vacher (1983). It assumed that the water table and the sharp interface maintain continuous equilibrium and that the salt water is static.

The two-fluid models

Alternatively, the two-fluid method may be used, in which coupled freshwater and salt water flow equations are solved simultaneously (e.g. Wilson and Sa da Costa, 1982; Contractor, 1983; Essaid, 1986; Willis and Finney, 1988). Most coupled two-fluid sharp interface models have been limited to a quasi-three-dimensional model single layer or a two-dimensional vertical section; however, Essaid (1990a,b) developed a quasi-three-dimensional model that allows for multiple aquifer layers. Saltwater dynamics can be important during the transient period; hence, a two-fluid model may be more appropriate for examining short-term responses (Essaid, 1986).

2.2 Solute Transport Models

When the problems involve miscible fluids, it is necessary to solve the solute transport equation. In order to solve the solute transport problem one has to solve the two equations: one governing equation of groundwater flow and another of solute transport equation.

2.2.1 The governing equation for solute transport without chemical reactions

$$\frac{\partial}{\partial x_i}\left(D_{ij}\frac{\partial c}{\partial x_j}\right)-\frac{\partial}{\partial x_i}\left(v_i c\right)=\frac{\partial c}{\partial t}, \qquad (2.3)$$

is also known as the advection-dispersion equation.
Where:

D_{ij} is the dispersion coefficient [L^2/T],

c is concentration[M/L^3],

v_i is the groundwater velocity [L/T], ($v_i = q_i/n$) ,

q_i is the specific volume flux and [L/T],

n is the porosity.

The code for a solute transport model typically consists of two submodels: a model to solve the flow equation and another to solve the advection-dispersion equation. The solution of the flow equation yields the distribution head, from which the velocity field is calculated. Velocities are input to the transport submodel, which predicts the concentration distribution in time and space. This holds true when the groundwater density is constant and it is also valid for water with low concentrations of total dissolved solids (TDS) and/or temperature in range of most shallow aquifers.

2.2.2 Density-dependent flow of miscible fluids

Simulation of flow involving water with high TDS or higher or lower temperatures requires that the effects of density be included in the model. This is the case of density dependent flow of miscible fluids that may be necessary to solve three models - flow, solute transport, and heat transport. Models that simulate density-dependent flow require initial pressure and density distribution. At the beginning of a time step, these initial values are used to generate the first approximation of the flow field. The resulting head values are input to the transport models, which redistribute solute and /or temperature. A new density distribution is calculated from the transport results, ending the first iteration of the first time step. The second iteration begins with the substitution of the newly calculated densities into the flow model. Iteration is continued until closure is attained. This process is repeated for all time steps (Anderson and Woesner, 1992).

2.2.3 Simulation models

Analytical models

The advection-dispersion equation can be solved analytically only after several simplifying assumptions e.g. a homogeneous aquifer and a uniform groundwater flow. The analytical solutions are obtained in either one-dimensional (Kreft and Zuber, 1978; Bear, 1979; Van Genuchten and Alves, 1982) or two-dimensional models of point injection (see Barends and Uffink, 1997).

Numerical models

Four major methods that solve the solute transport equation are: 1) the finite different method; 2) the finite element method; 3) the random walk method (Uffink, 1990); and 4) the method of characteristics (Konikow and Bredehoeft, 1978). In the last method, the particle tracking technique is also employed to solve the advective transport and either the finite difference or finite element approach is used to solve the dispersive equation.

2.3 Optimal saltwater intrusion management model under deterministic case

For optimal control of saltwater intrusion, the management model has been carried out under two forms. The first form is the so-called groundwater quality management, which uses the water quality (salt concentration in terms of Cl^-) as one criterion (objective function) for this type of management model (see Shamir et al., 1984; Oude Essink, 1996). The second one is more dominated by the salt/fresh water interface control management in which the location of the interface or the saltwater volume bounded by its interface are used as objective functions for the optimal management models (see Shamir et al., 1984; Willis and Finney, 1988; Finney et al., 1992; Emch and Yeh, 1998). In both cases, the cost objective function will always join in the total multi-objective formula of such problems.

For general groundwater quality management problems, Gorelick (1983) classified different types of management models into steady state and transient cases in linear programming management models as based on either the embedding method or the response matrix approach. Even though groundwater quality management models have been developed for those cases, research is still needed for the solution of non-linear groundwater quality control problems.

Non-linearities also arise in saltwater intrusion control problems. The difference in density between fresh and salt water serves as a significant driving force for the migration of solutes. In such cases, the groundwater velocity field is a function of solute concentrations. Hence non-linearities appear in advective and dispersive transport terms. Research is needed to develop distributed parameter management models of saltwater intrusion that involve simulation of this non-linear system (see Gorelick, 1983).

The conjunctive management of groundwater supplies and quality of regional aquifer systems is inherently a multi-objective planning problem, a problem characterized by conflicting objectives, constraints and policies (Willis and Yeh, 1987).

Various mathematical techniques have been developed to solve non-linear optimization problems. Quasi-linearization of any non-linearities within the objectives and constraint functions allows the application of linear programming methods (Emch and Yeh, 1998).

2.3.1 Non-linear multiple-optimization in saltwater intrusion management

The mathematical formulation of this multiple objective optimization problem can be stated as:

$$\text{Min } \{\boldsymbol{Z}(\boldsymbol{x}) = [Z_1(\boldsymbol{x}), Z_2(\boldsymbol{x}), \ldots, Z_k(\boldsymbol{x}), \ldots, Z_p(\boldsymbol{x})]\},$$

$$\text{subject to} \quad g_i(\boldsymbol{x}) \geq 0, \qquad i = 1, 2, \ldots, m,$$

$$x_j \geq 0, \;\; j = 1, 2, \ldots, n,$$

where x_j are decision variables of vector $\boldsymbol{x}$ for which optimal values are desired,

g_i are constraints,

Z_k is the kth objective function of vector $\boldsymbol{Z}$.

A. For the water quality management, a complex solute transport simulation model is combined with the non-linear optimization procedure (Gorelick, 1984). The aquifer management problem can then be expressed by at least one objective function of water quality criterion that might be written as follows: (see Shamir et al., 1984)

$$\text{Min } \{Z_k = \sum_i C_i\},$$

$$\text{subject to} \quad C_i \leq Cu_i,$$

where C_i is the Cl^- concentration in the cell i,

Cu_i is the upper bound of Cl^- concentration at cell i,

Z_k is the kth objective function of the vector $\boldsymbol{Z}$.

B. In the case of saltwater intrusion management problems based on the salt/ fresh water interface control model, Emch and Yeh (1998) presented a formulation of multi-objective non-linear programming under the set of n decision variables which is dependent on the number of wells, surface water sources, water users, and time periods for which optimal values are desired:

$$\boldsymbol{x} = \left(Q_{1,1}, \ldots, Q_{\omega,1}, Q_{1,2}, \ldots, Q_{\omega,2}, \ldots, Q_{i,j}, \ldots, Q_{\omega,\tau}\right),$$

where τ is the total number of time periods, $\omega = \omega_1 + \omega_2$ total number of supply sources; and $n = \omega \times \tau$.

the state variables are the fresh water heads, saltwater heads, and the interface position.

the two objectives are formulated as cost objective and saltwater intrusion (volume) objective.

Cost objective:

$$\text{Min } Z_1 = \sum_{j=1}^{\tau} \left[\sum_{i \in \Omega_1} Q_{i,j} \cdot C1_i \cdot \left(L_i - h_{i,j} \right) + \sum_{i \in \Omega_2} Q_{i,j} \cdot C2_i \right],$$

Saltwater intrusion (volume) objective:

$$\text{Min } Z_2 = \sum_{\ell \in \Gamma_1}^{\tau} \left[\iint z_I^{\ell} \left(x, y \right) dx dy \right],$$

where z_I^{ℓ} elevation of interface in layer ℓ [L],

L_i is the elevation of the ground surface above the datum at well i [L],

$h_{i,j}$ is the fresh water head at well i in period j [L],

$Q_{i,j}$ water supply rate from source i for time period j [L^3/T],

$C1_i$ unit cost for water extraction per height of required lift for source i [$\$/L^3/L$],

$C2_i$ unit cost for surface water supply for source i [$\$/L^3$].

each of these objectives is subject to the same set of constraints that may include supply source upper bounds and well capacity, demand, and drawdown constraints (formulated for time period $j = 1, \ldots, \tau$).

2.3.2 Linkage of simulation model to optimization model

Gorelick et al. (1984) linked the simulation model for flow and solute transport (SUTRA) with the optimization system (MINOS) as an independent module. They treated SUTRA as a subroutine that was called by the optimization procedure for function and Jacobian evaluation.

Finey et al. (1992) also linked MINOS to SHARP, and even more pertinently, they minimized the squared volume of saltwater in each aquifer of a layered system. In this study both the response matrix approach and an augmented Lagrangian method in conjunction with the reduced-gradient method were used.

Emch and Yeh (1998) also linked SHARP and MINOS with multi-objectives by incorporating SHARP into the optimization algorithm as a subroutine. Upon being provided new values of groundwater portion of the set of decision variables (pumping rates), SHARP returns head and interface position values. Subroutines within MINOS then calculate the appropriate objective or constraint function and finite difference techniques are used to determine the objective and constraint gradients of the management problem.

2.4 Groundwater quality management under uncertainty

One of the most difficult problems associated with the simulation-optimization approach to groundwater quality management is incorporating the effects of flow and transport modelling uncertainty into the optimal decision making process (Wagner and Gorelick, 1989). The uncertainty is due to the lack of knowledge concerning the spatial variability of the aquifer properties, mainly the spatial hydraulic conductivity. To date, the literature dealing with saltwater intrusion management models under uncertainty is unavailable whereas that dealing with groundwater quality management models under uncertainty is available by several authors e.g. Kaunas and Haimes (1985), Wagner and Gorelick (1986, 1987, 1989), Wagner (1988), Andrecevic and Kitanidis (1990), Wagner et al. (1992). A review of this literature can be found in the work by Wagner et al. (1992). In general, uncertainty has been dealt with in different ways in optimization models:

One traditional way of dealing with uncertainty in optimization models is to do post-optimality sensitivity analysis to determine the effect on the optimal solution of small changes in model data. This was done by Aguado et al. (1977), Willis (1979) and Gorelick (1982). See also Anderson and Woessner (1992).

Uncertainty can also be modelled using stochastic simulation. A method that has been used to incorporate uncertainty in the optimization model itself is to use chance constraints, so that certain constraints are not met exactly under all conditions, but instead are only met with a specified level of reliability (probability). This approach was developed by many authors e.g. Bredehoeft and Young (1983), Tung (1986), Wagner and Gorelick (1987, 1989), Hantush and Marino (1989), Maddock (1974), Ranjithan et al. (1990), and Andrecevic and Kitanidis (1990).

Risk analysis methods are also used to deal with uncertainty. In risk analysis the uncertainties in model inputs (such as timing and sizes of spills and leaks) are translated into uncertainties in outputs (such as probability of exceeding standards or the probability of contamination of a well). Risk analyses specifically dealing with groundwater contamination include Kaunas and Haimes (1985), Hobbs et al. (1988), and Lichtenberg et al. (1989).

Wagner and Gorelick (1989) used two main approaches in stochastic formulation that incorporate uncertainty into groundwater quality management models:

- The first approach, termed ***the multiple realization model***, is a non-linear simulation-optimization problem in which numerous realizations of the random hydraulic conductivity field are considered simultaneously. The solution of the multiple realization management problem is straightforward. That means that once the conditional hydraulic conductivity realizations are generated, the non-linear optimization problem is simultaneously solved for the N realizations. However the simultaneous solution of thousands of realizations (constraints) is simply not feasible. Therefore only 30 hydraulic conductivity realizations were put into the model. With this limitation, it cannot be assumed that the optimal management strategy is feasible for "all" possible conductivity fields or even for a high percentage of these fields. Therefore a post-optimality Monte Carlo analysis is performed to assess the reliability of the optimal solution.
- The second approach of the aquifer remediation problem in heterogeneous aquifers, which is called ***the Monte Carlo management model***, solves a series of individual optimization problems, each with a single realization of hydraulic

ERRATA

Pages	Lines	Errors	Corrections
29	3 from below	5.	2.
37	14	6.	1.
	15	7.	2.
	16	8.	3.
	17	9.	4.
	18	10.	5.
60	1	andZ_k	and Z_k
69	9 from below	11.	1.
	7 from below	12.	2.
	4 from below	13.	3.
	2 from below	14.	4.
70	1	15.	5.
	4	16.	6.
	6	17.	7.
	8	18.	8.
	10	19.	9.
	15	20.	10.
72	2 from below	21.	2.
	1 from below	22.	3.
73	6	23.	4.
	7	24.	5.
	8	25.	6.
	11	26.	7.
90	19 from below	2.	1.
	16 from below	27.	2.
	12 from below	28.	3.
	10 from below	29.	4.
97	10	3.	1.
	13	30.	2.
	15	31.	3.
	19	32.	4.
	21	33.	5.
	24	34.	6.
	28	35.	7.
	32	36.	8.
	38	37.	9.
118	4 from below	(5.35)	(5.63)

6. In een magnetisch veld is het niet gemakkelijk een stuk ijzer precies in het midden tussen de noord- en zuidpool van een magneet te houden. Evenzo zijn bij het besturen van de maatschappij de extreem linkse en rechtse standpunten niet beter maar wel gemakkelijker te verwezenlijken dan gematigde standpunten.

7. De gewoonte om te denken binnen een referentiekader met slechts absolute begrippen is de oorzaak dat sommige mensen extreme standpunten uitdragen.

8. Indien mensen, levend op het kritische niveau waar zij al optimaal presteren, toch iets meer willen bereiken moeten zij iets anders opgeven. Dit gebeurt meestal wanneer mensen denken dat zij meer kunnen dan hun vermogens hun toestaan omdat zij niet tevreden zijn met hetgeen zij verkrijgen met hun vermogens.

9. Het Nederlandse gezegde "Het Nederlandse weer is als een Nederlandse vrouw" en het Vietnamese gezegde "Vrouwen eerst en God als tweede" drukken de verschillende rollen uit van vrouwen in beide samenlevingen.

10. In het Sebastiaan Hostel (op de Zusterlaan, Delft) zou niet alleen de alarmbel maar ook noodinformatie en een telefoon in de lift geplaatst dienen te worden om mensen die vastraken te helpen in contact te treden met reddingswerkers.

STELLINGEN
behorend bij het proefschrift
"Op meervoudige doelstellingen gebaseerd beheer van zoutwaterintrusie in grondwater: optimalisatie en onzekerheid"
door
Tran Minh Thuan

1. Bij een optimimalisatie gebaseerd op meervoudige doelstellingen dienen de tegenstellingen tussen deze doelen zorgvuldig te worden gecontroleerd voordat men zo'n doelstelling van toepassing laat zijn op beheersproblemen. Dit komt voort uit het feit dat alle doelen niet altijd met elkaar in strijd zijn in elk stadium van beheer.

2. In het tweede stadium van het beheersprobleem (droge seizoen in de toepassing in dit proefschrift) kan de zoutwaterintrusie ten gevolge van de veranderende zoutwaterstromen als een grotere bedreiging worden gezien dan de zoutwaterintrusie ten gevolge van zoetwaterextractie.

3. Wanneer de zoetwaterinjectie moet stoppen ten gevolge van enig ongeval, dan zal het zoet-zoutscheidingsvlak (dat nog niet de stationaire situatie heeft bereikt) naderen tot bijna zijn beginpositie na hetzelfde tijdsinterval als nodig was om tot deze ligging te komen gedurende het injectieregiem.

4. De helling van de niet-inferieure verzameling (N_o) die gedefiniëerd is als de verhouding van de afname (ΔZ_2) van de zoutwaterintrusie tot de toename (ΔZ_1) van de kosten van injectie, wordt wel de "marginal rate of transformation" (MRT) genoemd. Voor het eerste stadium (natte seizoen in de toepassing in dit proefschrift) kan men de N_o opsplitsen in twee verzamelingen zodanig dat voor punten (Z_1,Z_2) in de eerste verzameling de MRT altijd groter is dan voor punten in de tweede verzameling. Dit weerspiegelt het feit dat toepassen van hogere kosten voor injectie op een gegeven moment minder effectief wordt.

5. Het is zeer moeilijk en kostbaar om zoutwaterindringing in grondwater in laag gelegen gebieden te voorkomen omdat hydrostatisch gezien de bewegende zoutwaterstroom nooit gelijktijdig met het stoppen van de kunstmatige zoetwateraanvulling de eindsituatie bereikt.

conductivity. Therefore if there are N hydraulic conductivity realizations, the Monte Carlo management model will provide n optimal reclamation strategies which each corresponds to a different realization of hydraulic conductivity. Since the hydraulic conductivity field is assumed to be random, the optimal pumping strategy is also random. Each of the n reclamation strategies obtained from the Monte Carlo management model represents a random sampling from the probability density function (*pdf*) of optimal pumping rates. Therefore the results of the Monte Carlo management model can be used to characterize the probability distribution of the optimal pumping rates.

Stochastic programming with recourse

Stochastic programming with recourse involves a two (or more) stage decision process. First a decision is made and implemented. At the later stage recourse action is taken, usually at some cost. Wagner et al. (1992) performed a stochastic programming with recourse for groundwater quality management. They considered the problem of containing an area of groundwater contamination by maintaining hydraulic gradients across a "capture curve" or "interception envelope." From this point of view, the groundwater system was modeled by embedding the discretizations of the partial differential equations governing groundwater flow as constraints in the optimization problem. The uncertainty in the values of hydraulic conductivity was taken into account. The recourse costs were modeled as a penalty that depends on the degree of "leakage" across the capture curve. This formulation is one of simple recourse, since the penalties are simply assessed, and are not a result of "second stage" decision made in order to minimize the recourse costs. An example of this formulation is given as follows:

A. Deterministic optimization problem:

The objective function: $$\text{Min} \sum_i A_1 w_i (s - h_i) - A_2 \left(\sum_i w_i \right)^2 ,$$

where

A_1 cost of pumping 1 m^3 water 1 m up, [$MUL^{-3}L^{-1}$], MU is a monetary unit.

w_i pumping rate in cell i, [L^3/T],

s height of the ground surface (measured from the bottom of the aquifer), [L];

h_i head in cell i, [L],

A_2 daily benefit, [$MUL^{-6}T$]; since the benefit term could be any linear or quadratic function of the pumping rate, w_i.

The constraints from embedding the descretization of the partial differential equations of groundwater flow are:

$$\sum_j F_{i,j} \, h_j = w_i - f_i \, , \qquad \forall i,$$

where $F_{i,j}$ are coefficients determining flow between cells i and j whose values depend on the geometry of the finite difference model and the hydraulic conductivities ($K_{x,y}$) and f_i are constant for cell i that depend on the boundary conditions.

Constraints are added to require that the head gradients (and thus the water flow) be inward across the capture curve:

$$h_I^{in} - h_I^{out} \leq 0 , \qquad \forall\, I ,$$

where h_I^{in} are the head values inside of the capture curve, h_I^{out} are the head values outside of the capture curve, and I is the index of cell pairs that form the boundary.

Constraints on the only pumping (no recharge) must be positive and below some maximum value:

$$0 \leq w_i \leq \overline{w} , \qquad \forall\, i .$$

Thus the deterministic optimization problem has a non-linear objective function subject to linear constraints.

B. Stochastic optimization problem:

The uncertainty in this problem is assumed to come from the stochastic nature of the hydraulic conductivities (K). Hence a set of realizations of the stochastic field of K is sampled, with each realization consisting of a distinct value for the hydraulic conductivity of each cell. These realizations are indexed by ω (ω = 1, 2,..., Ω). Each realization is assumed to occur with probability π_ω. Each realization ω will result in a different matrix F and, for a fixed pumping plan w, a different set of heads for each realization. Thus F_ω, h_ω, h_ω^{in}, h_ω^{out} are instead written in stochastic formulation as,

the objective function:

$$\text{Min} \sum_{\omega} \pi_\omega \left[\sum_i A_1 w_i \left(s - h_{i,\omega}\right) + \sum_{\ell} \rho\left(\upsilon_{\ell,\omega}\right) \right] - A_2 \left(\sum_i w_i \right)^2 ,$$

subject to: $\displaystyle\sum_j F_{i,j,\omega}\, h_{j,\omega} = w_i - f_i , \qquad \forall i, \ \forall \omega,$

$$\upsilon_{\ell,\omega} = h_{\ell,\omega}^{in} - h_{\ell,\omega}^{out} \leq 0 , \qquad \forall \ell , \quad \forall \omega ,$$

$$0 \leq w_i \leq \overline{w}, \qquad \forall\, i.$$

Where $\upsilon_{\ell,\omega}$ is the violation term, its positive value means that there is some leakage of contaminated water into the protected zone, past the plane capture curve. This leakage is assumed to result in a recourse cost, such as treatment costs for water pumped at the supply wells or costs for some other remedial action required to counteract the contamination of the protected zone. The recourse cost, ρ, is represented by a linear-quadratic penalty function for a violation, $\upsilon_{\ell,\omega}$, as follows:

$$\rho\left(\upsilon; p, q\right) = 0 , \qquad \upsilon \leq 0 ,$$

$$\rho\left(\upsilon; p, q\right) = \tfrac{1}{2} \upsilon^2 / p , \qquad 0 \leq \upsilon \leq pq ,$$

$$\rho\left(\upsilon; p, q\right) = q\upsilon - \tfrac{1}{2} pq^2 , \qquad \upsilon \geq pq .$$

The parameters p and q must be positive and are specified by the decision maker.

2.5 SOCO robust linear programming

A new approach for the stochastic optimization that has been applied to many aspects of engineering fields is the quadratic cone or so-called second order cone optimization (SOCO) programming (Ben-Tal and Nemirovski, 1998). This approach has been applied for the first time to the multi-objective groundwater quantity management (for the confined aquifer only) by Ndambuki, 2001. In his problem the Modflow is linked with SeDuMi (Sturm, 1998-2002) by the hydraulic head response matrix approach. Instead of determining the first stage optimal solution for a robust optimal solution, the new approach will treat the uncertainty problem under the robust least-squared method (see Ghaoui and Lebret, 1997). In this approach the uncertainty in input parameters will be transformed into the uncertainty ellipsoids whose centres are the mean values (nominal values) and the deviations from any points within the ellipsoids to the centers represent the uncertainty of the input parameters (perturbations). This approach is a promising tool for a large-scale optimization problem in the sense of less CPU-time compared to the stochastic optimization with recourse (see more details in Ndambuki, 2001). In fact Ndambuki (2001) carried out his work by even the multiple scenario scheme (a latest variant of the classical Monte Carlo approach), but the question of which optimal strategy to choose among the many scenarios for implementation is not very obvious (Ndambuki, 2001). Moreover, the optimal strategy corresponding to, for instance, a worst-case scenario will not guarantee to satisfy the constraints for all the other realizations (Ndambuki, 2001). Hence, this limited the robustness of the stochastic optimization problem. For our stochastic SW intrusion management problem, which has more complexities, applying this new approach, SOCO robust linear programming, will be more appropriate than other approaches in terms of increasing the robustness while considerably reducing the CPU time consumption.

2.6 Conclusions

Summarizing the above discussion we can conclude that:

- Strictly speaking, the saltwater intrusion simulation should be mathematically based on the whole of governing equations of fluid flow and salt mass transport that obey the principle of mass conservation. However many problems show that the dynamics of the zone of contact between fresh water and saltwater, either considered as a sharp interface or a dispersive mixing zone, play a key role in understanding and managing practical seawater intrusion problems. Models that incorporate simulation of the transition zone may require simultaneous solution of those two equations. Three dimensional density-dependent flow and solute transport codes have been developed but the increased computation effort required to solve them has limited most solutions to two-dimensional vertical cross sections. In cases where the transition zone is very dispersed and chloride concentration gradients are low the effects of variable density may be neglected, allowing decoupling of the governing equations (Emch and Yeh, 1998). The sharp interface approach, in conjunction with integration of the flow equations (fresh and saltwater flow) over the vertical (Essaid, 1990), can be applied regionally to large physical systems. This approach does not give information concerning the nature of the transition zone; however, it does reproduce the regional flow dynamics of the system and response of the interface to applied stresses. Volker and Rushton, 1982, compared steady state solutions for both the disperse and sharp interface approaches and showed that as the coefficient of dispersion decreases the two solutions approach each other. The sharp interface models which simulate flow only in the freshwater region, by incorporating the Ghyben-Herzberg approximation, assume that the saltwater domain adjusts rapidly to applied stresses. In many cases, to reproduce the short-term behavior of a coastal aquifer, it is necessary to include the influence of saltwater flow (Essaid, 1986). This two-fluid flow model will be applied in the next chapter by a sensitivity analysis.

- Management of coastal aquifer reclamation is guided by several criteria: groundwater levels, location of the interface, salt concentration, saltwater intrusion volume, the costs of pumping and recharge, etc. In this work we want to study the SW intrusion length that is defined by the distance of the location of interface toe from the shoreline. Combining the two powerful analysis techniques, simulation and optimization, produces an engineering design tool, which can aid in the formulation of design criteria and assists decision makers in assessing the impacts of design trade-offs. Management of saltwater intrusion problems often requires the use of non-linear optimization models. This is due to the complexity of the governing equations and the non-linear response of the tracer concentration or the sharp interface with respect to applied stresses. Solution techniques can be classified as either unconstrained, linearly constrained or non-linearly constrained optimization methods. This will be discussed in Chapter 5.

- Stochastic programming for the problem of saltwater intrusion management under uncertainty is also necessary. This is the first time in the literature that uncertainty has been introduced into such a problem. Since the coastal aquifer management models used to simulate and design for the optimal control of flow and salt mass transport have been assumed to be deterministic, this method was used prevalently by Shamir et al. (1984), Willis and Finney (1988), Finney et al. (1992) and Emch and Yeh (1998). The model parameters that govern groundwater flow and contaminant transport are assumed to be precisely known. Unfortunately, we never know the precise values of the model parameters. They are always estimated, commonly with any of a number of inverse techniques, using (imprecise) data collected in the laboratory and/or the field. Most of the information sources have been based on the surveying and monitoring of aquifers and salinity of groundwater. Unfortunately, good data is often the weakest point of many actual studies. Therefore there is a degree of uncertainty associated with the parameters used to simulate aquifer behavior and, consequently, the simulated tracer concentration/fluid densities or salt/fresh water interface and pressures or hydraulic heads are themselves uncertain. This problem with the salt/fresh water interface and the hydraulic heads will be discussed and solved with the quadratic cone programming in Chapters 5, 6 and 7.

Chapter 3

The responses of saltwater intrusion lengths with respect to the stresses and transmissivities - Sensitivity analysis based on the SHARP computer code

3.1 Introduction

An essential step in modeling applications is the sensitivity analysis to quantify the uncertainty in the model caused by lack of data or uncertain aquifer parameters, stresses, and boundary conditions. Besides the heads, it is also important to know the sensitivity of the intrusion lengths depending on these parameters. Moreover, the knowledge of sensitivity analysis of heads and intrusion lengths can be used during operational management later on.

The model application has been hypothetically made for the Ben Tre aquifer system, which is located on the Southeast of the Mekong Delta (MD) of Viet Nam. Reportedly, this aquifer is partly intruded by seawater (see Michael, 1971). In a sense of conjunctive use of surface water and groundwater for the regional development prospect, the increase of recharge and a groundwater barrier against seawater intrusion may be necessarily one of alternatives.

3.2 The governing equations of the two-fluid flow model (SHARP)

The governing equations describing two-fluid flow, which is considered as the movement of an immiscible fluid, can be written according to Bear (1979), and Essaid (1986):

$$\left[S_f b_f + \phi(a+\delta)\right]\frac{\partial h_f}{\partial t} - \phi(1+\delta)\frac{\partial h_s}{\partial t} = \frac{\partial}{\partial x}\left(b_f K_{fx}\frac{\partial h_f}{\partial x}\right) - \frac{\partial}{\partial y}\left(b_f K_{fy}\frac{\partial h_f}{\partial y}\right) + Q_f + R_f, \tag{3.1.a}$$

and

$$\left[S_s b_s + \phi(1+\delta)\right]\frac{\partial h_s}{\partial t} - \phi\delta\frac{\partial h_s}{\partial t} = \frac{\partial}{\partial x}\left(b_s K_{sx}\frac{\partial h_s}{\partial x}\right) - \frac{\partial}{\partial y}\left(b_s K_{sy}\frac{\partial h_s}{\partial y}\right) + Q_s + R_s, \tag{3.1.b}$$

with the interface elevation can be calculated from the fresh and saltwater heads:

$$z_i = (1+\delta)h_s - \delta h_f, \tag{3.1.c}$$

where

h_f, h_s	are the freshwater head and saltwater hydraulic head [L],
S_f, S_s	freshwater and saltwater specific storage [1/L],
b_f, b_s	average saturation thickness of freshwater and saltwater zones [L],
ϕ	the effective porosity,
δ	$= \gamma_f / (\gamma_s - \gamma_f)$,
γ_f, γ_s	freshwater and saltwater specific weight [M L^{-2} T^{-2}],

R_f, R_s	fresh and saltwater leakage across top and bottom of aquifer [L/T],
Q_f, Q_s	fresh and saltwater source/sink terms (pumping, recharge) rates [L/T],
K_{fx}, K_{fy}	freshwater hydraulic conductivities [L/T],
K_{sx}, K_{sy}	saltwater hydraulic conductivities [L/T],
a	$a = 1$ if the aquifer is unconfined; $a = 0$ if the aquifer is confined.

In order to solve the dynamic flow of saltwater intrusion, the coupled freshwater and saltwater flow equations are solved simultaneously. The finite-different approximation has been used to discretize the differential equations and solve by strongly implicit procedure (SIP) in order to obtain the iterative solution (see Essaid, 1990).

3.3 Quasi-two-dimensional model

Here we will simulate the quasi-two-dimensional confined aquifer case by applying the SHARP model, a quasi-three-dimensional model (Essaid, 1990). In the horizontal plane there are three rows of 1250m each and 330 columns of 250m each. The hydrogeological data and aquifer parameters will be as follows:

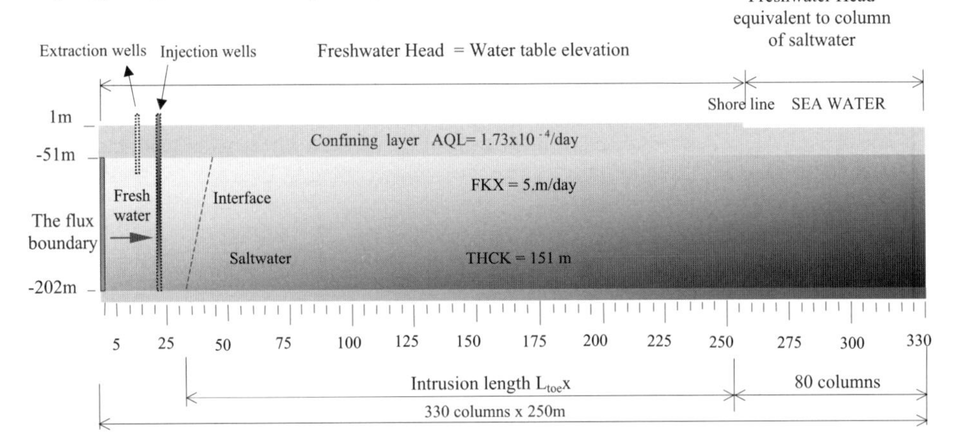

Figure 3.1: Cross sectional profile of the model.

The topmost one is the semi pervious Holocene layer, with C = 5780 days. The permeable Pleistocene layer (main aquifer), which is assumed to be homogenous, has a hydraulic conductivity of 5 m/day and thickness is 151m.

The bottom of the system is considered to be impermeable, i.e. a no-flow boundary. A constant flux boundary is considered at the left boundary, with values varying in the range of 864 m^3/d – 43200 m^3/d. A partly penetrating extraction and a fully penetrating injection well are located at cells 12 and 14, respectively. The geometry of this problem is illustrated in Figure 3.1.

3.4 Quasi-three-dimensional model

The model consists of a rectangular area in the horizontal plane. Its aquifer system is considered as confined, having the same cross sectional profile as in the quasi-two-dimensional model (see Figure 3.1).

In the horizontal plane view, we simulate the quasi-three-dimensional grid system in which there are 15 rows of 1250m each in the y-direction. Grid spacing in the x-direction is 1250m except near the location of wells and the interface where it is reduced to 250m to improve the interface projection accuracy (see Essaid, 1990). Therefore the total number of columns in x-direction is 146. The salt/fresh interface is considered as an abrupt change of the groundwater density, implying no existence of the mixing zone of brackish water. The hydro-geological data and aquifer parameters will be the same as in the first case.

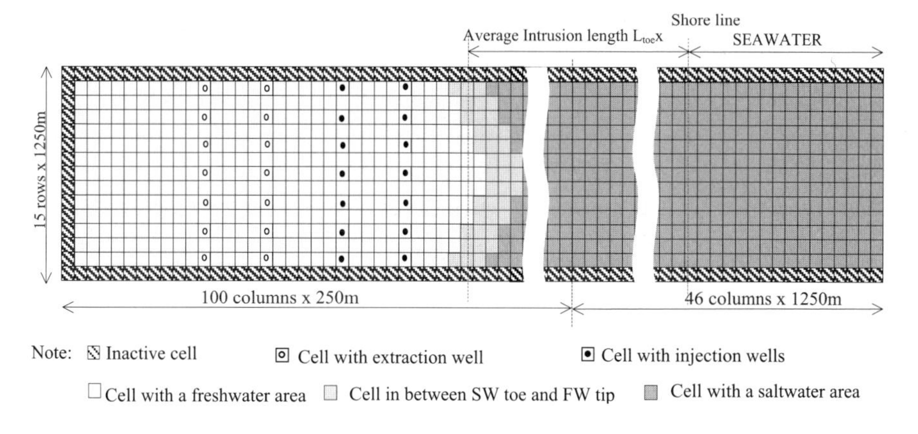

Figure 3.2: The horizontal plane view of the aquifer of the quasi-three-dimension model.

3.5 Objectives and Procedures

3.5.1 Quasi-two-dimensional model

For the quasi-two-dimensional model, the sensitivity analysis is focussed on:

1. Influence of boundary flow rates on the responses of intrusion lengths and freshwater heads;
2. Responses of the intrusion lengths to the injection rates;
3. Responses of the intrusion lengths to the extraction rates; and
4. Responses with respect to the changes of the aquifer hydraulic conductivity.

3.5.2 Quasi-three-dimensional model

1. The sensitivity analysis aims at understanding of the behaviour of the intrusion length in the horizontal plane. It enables incorporation of the simulation model into the optimization problem by taking the gradients of the response curves of the horizontally averaged intrusion toe length with respect to the injection/extraction rate perturbations.
5. Consequently, the sensitivity analysis here is performed on the responses of the average intrusion toe length to the injection or extraction rates perturbations in turn among a set of 28 active well locations in the model area.

3.6 Model Simulation Results

The model simulation results for this sensitivity analysis are obtained when the steady state is finally achieved. (Please note that the seaside in Figures 3.3, 3.6, 3.9 and 3.12 is to the left while the seaside in Figures 3.1 and 3.2 is to the right)

3.6.1 The Flux Boundary Rate Perturbations

This is a case of the new run, the initial interface elevations are calculated from the Ghyben-Herzberg relation, $\zeta_1 = -\delta\phi_f$, where $\delta = \gamma_f / (\gamma_s - \gamma_f)$.

A freshwater and saltwater specific storage of 0.0 is used in order to accelerate the steady state achievement. Figure 3.3 shows the responses of both the hydraulic heads and the sharp interface with respect to the flux boundary. When the flow rates at the

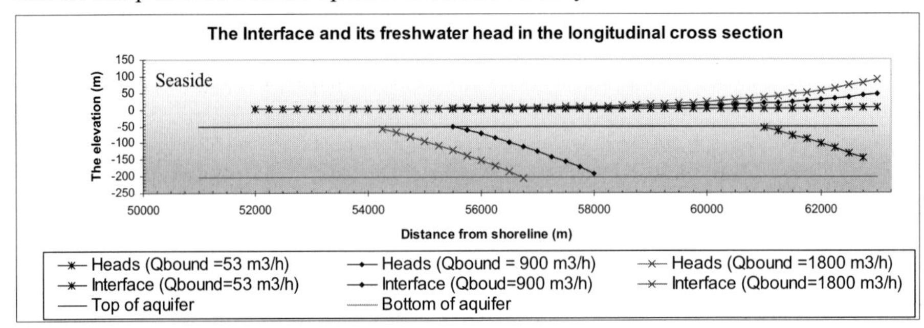

Figure 3.3: The responses of the freshwater head and interface to the boundary flow rates.

boundary vary from small to big values the freshwater head at a cell is proportionally increased while the interface moves seaward with the non-linear inverse proportion to these flow rates.

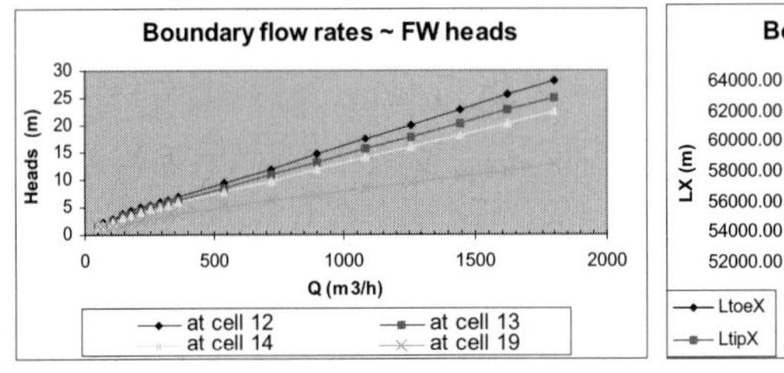

Figure 3.4: Freshwater heads versus boundary flow rates

Figure 3.5: Intrusion lengths versus boundary flow rates

Graphically, the relationship of the freshwater head at a cell and the boundary flow rate is represented by a straight line, as shown in Figure 3.4. In contrast, the response of the saltwater intrusion lengths (tip and toe) with respect to these flow rates are of curvilinear relationships, as shown in Figure 3.5.

3.6.2 The Injection Rate Perturbations

This is the case of a continuation run, where the specific storage is chosen as 2.0 10^{-4} m^{-1}. The initial interface *ZINT*, the interface tip projection in x and y direction *FX*, *FY* and the interface toe projection in x and y direction *SX*, *SY* are taken from the results of the boundary flow case with the rate of $1800 m^3/h$. Here the terms of tip or toe projection in x and y direction are defined as the distance to the interface tip or toe in block i, j expressed as a ratio of Δx and Δy respectively.

- A constant boundary flow of 1800m^3/h ($\approx$ 0.5m^3/s) is put on the upstream boundary. The boundary flow together with the initial freshwater head ensures that the freshwater zone exists in the aquifer at the beginning of the continuation run.

- The extraction rate is taken at a constant of 36 m^3/h ($\approx$ 0.01m^3/s) at cell (column) 12.

- Then a number of the injection rates varying from 0 to 1800 m^3/h ($\approx$ 0.5m^3/s) at cell (column) 14 will be applied to the injection cell in order to observe the responses of the freshwater heads and the sharp interface.

The steady state is achieved for all perturbation runs.

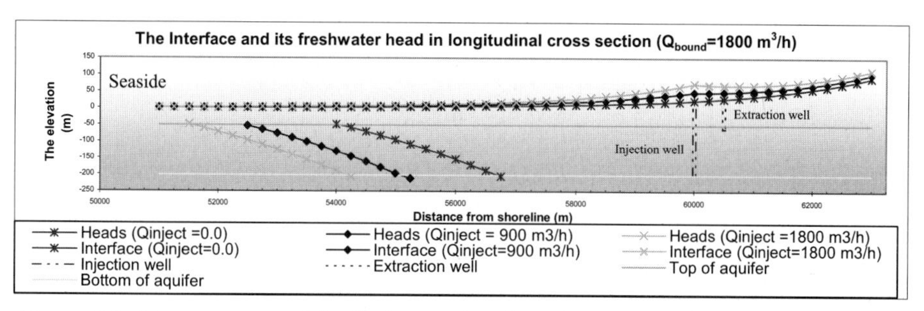

Figure 3.6: The responses of the freshwater head and interface to the injection rates.

The results in Figure 3.6 shows that the responses of freshwater heads and the interface with respect to the injection rates have the same behaviour as in the boundary flow rate case. This is because when the injection (or the boundary flow) rate increases, the freshwater head linearly increases and the intrusion lengths decrease non-linearly. Hence, the relationships of the freshwater head at a cell and the intrusion lengths versus the injection rates are also linear and non-linear, respectively shown in Figure 3.7 and Figure 3.8.

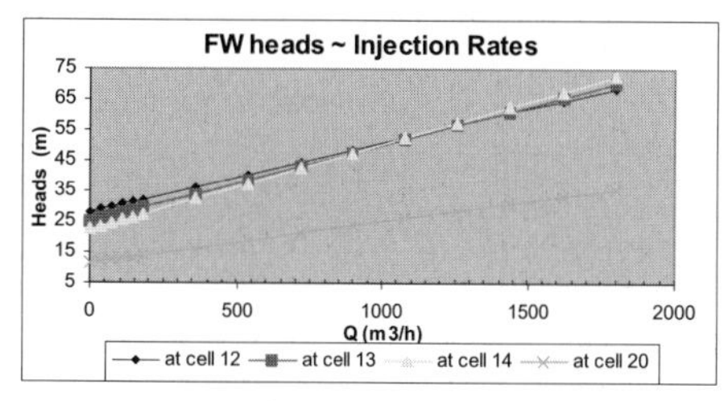

Figure 3.7: Freshwater heads versus injection rates

Figure 3.8: Intrusion lengths versus injection rates

3.6.3 The Extraction Rate Perturbations

This is also the case with the continuation run, where the specific storage is chosen as 2.0 10^{-4}m^{-1} and the initial interfaces *ZINT*, *FX*, *FY*, *SX*, *SY* are taken from the results of the injection case with the injection rate of 1800 m^3/h ($\approx$ 0.5 m^3/s).

- The same constant flux boundary of 1800 m^3/h ($\approx$ 0.5 m^3/s) is also put on the upstream boundary.

- The injection rate is taken constant of 1800 m^3/h ($\approx$ 0.5 m^3/s) at cell (column) 14. It ensures that the interface must be as far as possible from the extraction well (or from the left boundary). In this case the initial distance of the interface toe from the shoreline is 54792.25m at cell 34.

- The number of extraction rates varying from 36 m^3/h - 2340 m^3/h ($\approx$ 0.01 m^3/s - 0.65 m^3/s) will be applied to the extraction cell (column 12).

- The steady state is also achieved for all perturbation runs.

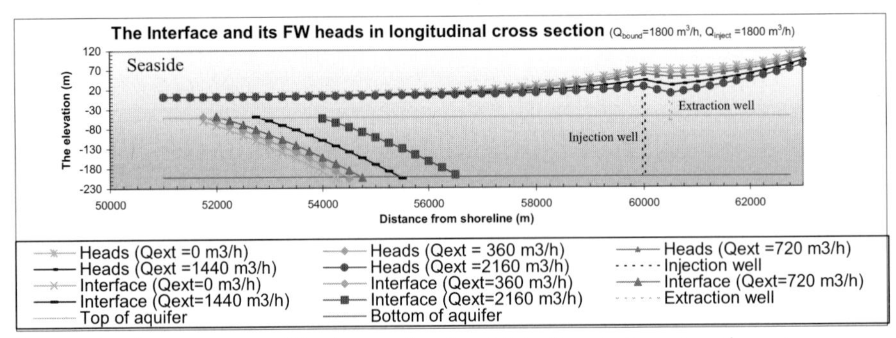

Figure 3.9: The responses of freshwater head and interface to the extraction rates.

The graphical relationships of the freshwater head and the intrusion lengths with respect to extraction rates are also linear and curvilinear, respectively. However, in this case these relationships are inverse to the case of injection rate perturbations. This means that when the extraction increases, the freshwater head linearly decreases, and the intrusion lengths non-linearly increase, as shown in Figures 3.10 and 3.11.

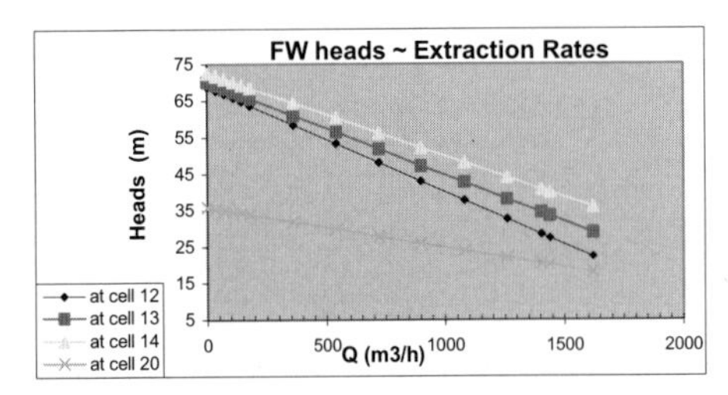

Figure 3.10: Freshwater heads versus extraction rates

Figure 3.11: Intrusion lengths versus extraction rates

3.6.4 Variations of Hydraulic Conductivity

The aquifer of all the following experimental cases is assumed to be homogeneous. A number of hydraulic conductivity values are chosen within a range from 0.5 m/d to 40 m/d for all cases. The constant flow rates of 1800 m^3/h, 1800 m^3/h and 2340 m^3/h are chosen for the boundary flow, injection and extraction rates, respectively.

The results of this case are as follows:

- The absolute values of head gradients in cases of low hydraulic conductivity are also greater than the ones in cases of high hydraulic conductivity especially at the extraction/injection well locations, see Figure 3.12.

- The interface slope (tgα or α) always decreases when the hydraulic conductivity value increases, see Figures 3.12 and 3.13.

- The results also show that while the hydraulic conductivity increases, the freshwater tip intrusion length always decreases, whereas the saltwater toe intrusion length does not always increase. See Figure 3.14.

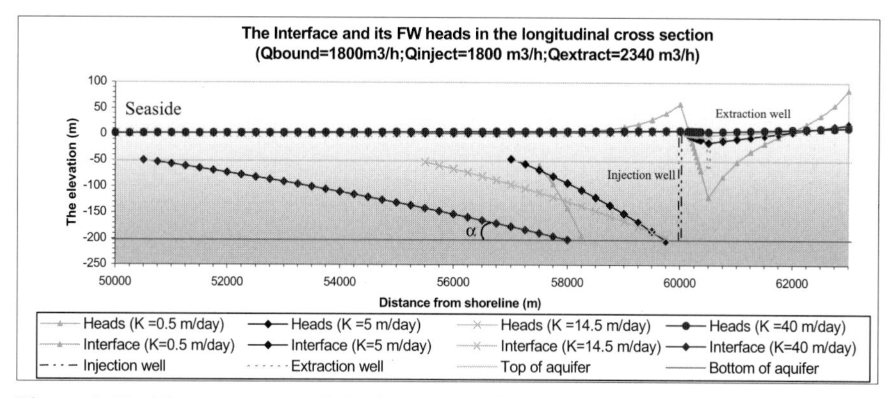

Figure 3.12: The responses of freshwater head and interface to the changes of the hydraulic conductivity.

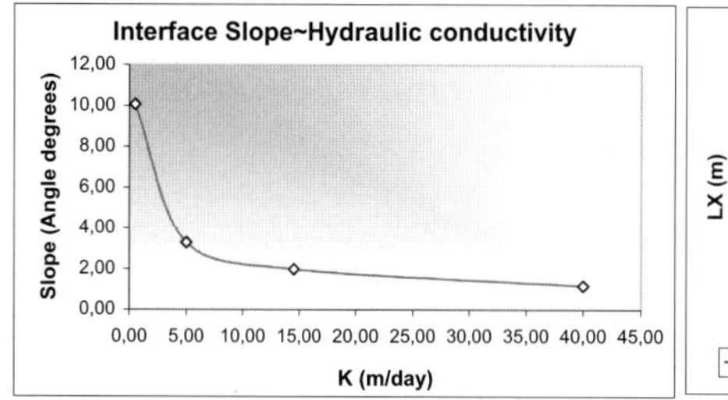

Figure 3.13: Interface slope versus hydraulic conductivity

Figure 3.14: Intrusion lengths versus hydraulic conductivity

3.6.5 The average intrusion increments with respect to stresses in the quasi-three-dimensional model

In the quasi-three-dimensional model, the stresses are perturbed in the range of 0 m^3/h – 900 m^3/h. The model runs separately for the twenty-eight single wells. Each single well is subjected to such a series of perturbed stresses. The responses of the interface toe with respect to such stresses are reported under the form of either the increment or the decrement of the averaged intrusion length of the saltwater toe. The average intrusion length is compared to the initial intrusion length of the interface toe at the non-pumping condition.

The numerical results show that the relationship between the average intrusion toe lengths and the extraction/injection rates is also of a non-linear nature. At a same stress value, the closer the location of the well is to the interface, the bigger the gradient of its curve is. With respect to a small range of stress values, the response-curved gradients of wells that are close to the boundaries (in *y*-direction) are bigger than the gradients of wells that are in the middle for the case of injection. It inversely holds for the case of extraction.

These results are shown in Figure 3.15.

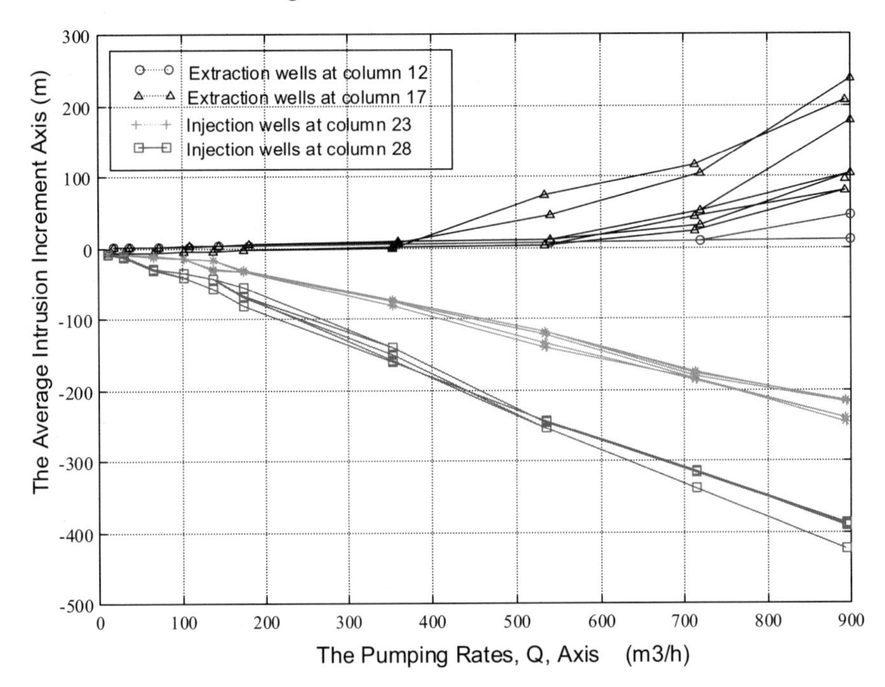

Figure 3.15: The average intrusion increments and pumping rates, Q, in the quasi-three-dimensional model. (The wells at columns 12, 17, 23, 28 are far from the shoreline with the distances of 59500 m, 58250 m, 56750 m and 55500 m respectively.)

3.6.6 Results and discussion

For the quasi-two-dimensional case:

- The responses of freshwater (FW) head in a control point are linear with respect to the changes of boundary flow rates, injection rates or extraction rates at a certain well location.
- The movement of the interface (or the intrusion lengths) is non-linear with respect to linear changes of boundary flow rates, injection rates or extraction rates at a certain well location. In the case of injection, the gradients of its curve decrease when stress is increasing. For the extraction case, the curvilinear relationship is reversed implying that its tangents will vary from flat to steep.

- The slopes of interfaces of all cases have more or less the same value (with α = $3^{o}28'$ -$3^{o}29'$) when the hydraulic conductivity is kept constant (and equal to 5m/day).
- The interface slope changes considerably from steep to flat if the hydraulic conductivity varies from low to high values. The absolute values of head gradients in cases of low hydraulic conductivity are also greater than the ones in cases of high hydraulic conductivity, especially near the flow boundary or extraction/injection well locations.

For the quasi-three-dimensional case:

- The relationship between the average intrusion toe lengths and the extraction/injection rates is also of a non-linear nature.
- At the same stress value, the closer the location of the well is to the interface, the bigger the gradient of its curve.
- In a small range of stress value, the response-curved gradients of wells, which are close to the boundaries (in the y-direction) are bigger than the ones of wells which are in the middle in case of injection. It holds inversely in the case of extraction.

3.7 Conclusions

- In most cases, the model results are quite dependent on the initial freshwater heads, which must be given to each of the injection wells in case the initial interface has to be calculated by the Ghyben-Herzberg formula.
- The results of the sensitivity analysis are going to be incorporated for the deterministic linear optimization problem (see Sturm, 2001) in Chapters 5, 6 and 7.
- The perturbation size of the stress selected should be as small as possible, such that the response increments of the averaged intrusion lengths are still greater than the rounding errors of the computation (Ahlfeld et al., 2000). This can be achieved using an iterative method to ensure convergence.
- The different responses of the interface with respect to the variation of the hydraulic conductivity will be extended to the heterogeneous case, which can be incorporated into a robust linear programming and solved by the quadratic cone (second-order cone) optimization problem (Ben-Tal et al., 1998; Ndambuki, 2001).

Chapter 4

Conic Quadratic Programming

In this chapter the concept of conic quadratic programming (see Ben-Tal and Nemirovski, 2001) will be reviewed that will be applied later to our optimization problems. Its principle will be introduced through the general concept of conic programming that is based on the cone set's theory for linear programming. The following sections, which are a summary of the lectures by Ben-Tal and Nemirovski (see Ben-Tal and Nemirovski, 2001), may hopefully help the reader to make acquaintance with quadratic cone programming.

4.1 Linear programming

The simplest type of the problems in management and planning that one usually sees are in the form of linear programming. This type of problem is defined by the property that the objectives and all inequality or equality constraints are linear functions. In order to solve these problems there are methods available i.e., the simplex method and its variants (for the linear case only) and the interior point method that has been predominantly studied nowadays by many researchers.

4.1.1 A primal form of linear programming

We know that a linear programming (LP) program is an optimization program that has the general form:

$$\text{Min } c^T x,$$

subject to

$$Ax \geq b, x \in \mathbf{R}^n .$$

If we consider this form as the original form of the problem, it is called the primal form of the linear programming problem.

It can be rewritten in another form:

$$c^T x \rightarrow \text{Min } \left| Ax - b \geq 0 \right. \qquad \text{(LP)}$$

where

$x \in \mathbf{R}^n$ is the design vector (its components are called decision variables)

$c \in \mathbf{R}^n$ is a vector of coefficients given to the objective

A is a given $m \times n$ constraint matrix, and $b \in \mathbf{R}^m$ is a given right hand side of the inequality constraints.

The (LP) is called (see Ben-Tal and Nemirovski, 2001, p. 1-2):

- feasible if its feasible set $\{x | Ax - b \geq 0\}$ is nonempty; a point from this nonempty set is called a feasible solution to (LP);
- bounded below, if it is either infeasible, or its objective $c^T x$ is bounded below on the feasible set.

It is convenient to use the notation of the largest lower bound of the objective on the feasible set as

$$c^* \equiv \inf_{x:Ax-b\geq 0} c^T x$$

to indicate the optimal value of a feasible below bounded problem.

4.1.2 A dual form of linear programming

Now if we consider the (LP) as a primal form of itself, then it has its dual form as below (see Ben-Tal and Nemirovski, 2001, p. 16-17):

$$b^T y \rightarrow \text{Max} \left| A^T y = c, y \geq 0 \right. \qquad \text{(LP*)}$$

where y is a feasible solution to (LP*) if (LP) is feasible, with the corresponding value of the dual objective $b^T y$ being a lower bound on the optimal value c^* in (LP); or vice versa, for every lower bound a on the optimal value of (LP), there exist a feasible solution y to (LP*) with $b^T y \geq a$.

According to the LP duality theorem in linear programming (Ben-Tal and Nemirovski, 2001, p. 18) we have the following five properties that are equivalent to each other:

6. The primal is feasible and below bounded.
7. The dual is feasible and above bounded.
8. The primal is solvable.
9. The dual is solvable.
10. Both primal and dual are feasible.

Also from the theorem (Ben-Tal and Nemirovski, 2001, p. 19), we can consider an LP program (LP) along with its dual (LP*), and let (x, y) be a pair of primal and dual feasible solutions. The pair is comprised of optimal solutions to the respective problems if and only if

$$y_i [Ax - b]_i = 0, \quad i = 1,...,m, \qquad \text{[complementary slackness]}$$

as well as if, and only if,

$$c^T x - b^T y = 0 \ . \qquad \text{[zero duality gap]}$$

4.2 From linear programming to conic programming

The linear programming is in the class of convex programming since its objectives and constraints are based on linear functions and the feasible set is assumed to be a convex set.

4.2.1 Orderings of $\mathbf{R}^m$

The constraint inequality $Ax \geq b$ in (LP) is an inequality between vectors. In general we can compare all respective components of two vectors by defining the ordering between the two vectors. Given two vectors $a, b \in \mathbf{R}^m$, and $a \geq b$ if the coordinates of a majorate the corresponding coordinates of b, we can write

$$a \geq b \Leftrightarrow \{a_i \geq b_i, \quad i = 1,...,m\},$$

where in the relation $a_i \geq b_i$ the sign " $\geq$" is an arithmetic $\geq$ between real numbers.

The inequality that is understood as a "coordinate-wise" partial ordering of vectors from $\mathbf{R}^m$ satisfies a number of basic properties of the standard ordering of reals; namely, for all vectors $a, b, c, d,\ldots \in \mathbf{R}^m$ one has

1. Reflexivity: $a \geq a$;
2. Anti-symmetry: if both $a \geq b$ and $b \geq a$, then $a = b$;
3. Transitivity: if both $a \geq b$ and $b \geq c$, then $a \geq c$;
4. Compatibility with linear operations:
   (a) Homogeneity: if $a \geq b$ and λ is a nonnegative real, then $\lambda a \geq \lambda b$,
   (b) Additivity: if both $a \geq b$ and $c \geq d$, then $a+c \geq b+d$.

We can see that a significant part of the useful features of LP programs comes from the fact that the vector inequality in the constraint of (LP) satisfies the properties (axioms) 1,…, 4.

4.2.2 Conic set - Convex cones

By the cone theory we have the definition of cones and convex cones as follows (see Steuer, 1986):

Cones: Let $v\in V\subset \mathbf{R}^m$, $V\neq \varnothing$. Then, V is a cone if, and only if, $\lambda v \in V$ for all scalars $\lambda \geq 0$. The origin $\mathbf{0} \in \mathbf{R}^m$ is contained in every cone. A cone need not be convex.

Convex cones: If the cone is considered as a set, a so-called conic set, it can inherit the definition of the convex set and be defined as: a cone $V \subset \mathbf{R}^m$ is convex if, and only if, for any vectors v^1, $v^2\in V$, the vector $(\lambda v^1+(1-\lambda)v^2) \in V$ for all $\lambda \in [0,1]$.

Pointed cones: A cone that has an extreme point (vertex) is said to be a pointed cone. The hyperplane $\{x\in \mathbf{R}^m \,|\, c^T x = 0,\ c\in \mathbf{R}^m\}$ and the closed half-space $\{x\in \mathbf{R}^m \,|\, c^T x \geq 0,\ c\in \mathbf{R}^m\}$ are nonpointed cones.

By the above definitions, the nonnegative orthant $\mathbf{R}^m{}_+$ is a pointed convex cone and $\mathbf{R}^m$ is a convex cone but it is not a pointed cone.

Now we consider vectors from $\mathbf{R}^m$ and assume that this set is equipped with a partial ordering, let it be denoted by $\succeq$. What happens when the pair of vectors $a, b \in \mathbf{R}^m$ is linked by the inequality $a\succeq b$. When we say that our ordering is "good", we mean that it fits the axioms 1,...,4.

C. A good inequality $\succeq$ is completely identified by the set $\mathbf{K}$ of $\succeq$-nonnegative vectors:

$$\mathbf{K} = \{a\in \mathbf{R}^n \,|\, a \succeq 0\}.$$

For instance, if $a\succeq b$, by axiom 1, we have $-b \succeq -b$ then

$$a \succeq b \Leftrightarrow a-b \succeq 0 [\Leftrightarrow a-b\in \mathbf{K}].$$

This set however cannot be arbitrary:

D. In order for a set $\mathbf{K}\subset \mathbf{R}^m$ to define, via the rule

$$a \succeq b \Leftrightarrow a-b\in K \qquad (*)$$

a good partial ordering on $\mathbf{R}^m$, it is necessary and sufficient for $\mathbf{K}$ to be a pointed convex cone, i.e., to satisfy the following conditions:

1. $\mathbf{K}$ is nonempty and closed with respect to addition of its elements:

$$a, a' \in \mathbf{K} \Leftrightarrow a + a' \in \mathbf{K}$$

(since if $a \succeq 0$ and $a' \succeq 0 \Leftrightarrow a + a' \succeq 0 \Leftrightarrow a + a' \in \mathbf{K}$).

2. $\mathbf{K}$ is a conic set – along with all points a, it contains the ray $\{\lambda a | \lambda \geq 0\}$.

3. $\mathbf{K}$ is pointed – the only vector a such that both a and $-a$ belong to $\mathbf{K}$ is the zero vector.

$$\text{(if } a \succeq 0 \text{ and } -a \succeq -a \Leftrightarrow a - a \succeq 0 - a \Leftrightarrow 0 \succeq -a \Leftrightarrow \begin{cases} -a \notin \mathbf{K}, \text{ if } a \succ 0 \\ -a \in \mathbf{K}, \text{ if } a = 0 = -a \end{cases}.$$

Thus, every pointed convex cone $\mathbf{K}$ in $\mathbf{R}^m$ defines, via the rule (*), a partial ordering on $\mathbf{R}^m$ which satisfies the axiom 1,…, 4. This ordering is denoted by $\geq_{\mathbf{K}}$:

$$a \geq_{\mathbf{K}} b \Leftrightarrow a - b \geq_{\mathbf{K}} 0 \Leftrightarrow a - b \in \mathbf{K}$$

We can see clearly that if $\mathbf{K}$ is the cone containing the vectors with nonnegative components - the nonnegative orthant, $\mathbf{R}^m_+$:

$$\mathbf{R}^m_+ = \left\{ x = (x_1, \ldots, x_m)^T \in \mathbf{R}^m : x_i \geq 0, i = 1, \ldots, m \right\}.$$

The Lorentz (or second order; less scientific name: the ice-cream) cone is an example of other partial orderings (shown in Figure 4.1):

$$x \geq_{\mathbf{L}^m} 0 \Leftrightarrow x \in \mathbf{L}^m : \mathbf{L}^m = \left\{ x = (x_1, \ldots, x_{m-1}, x_m)^T \in \mathbf{R}^m : x_m \geq \sqrt{\sum_{i=1}^{m-1} x_i^2} \right\}.$$

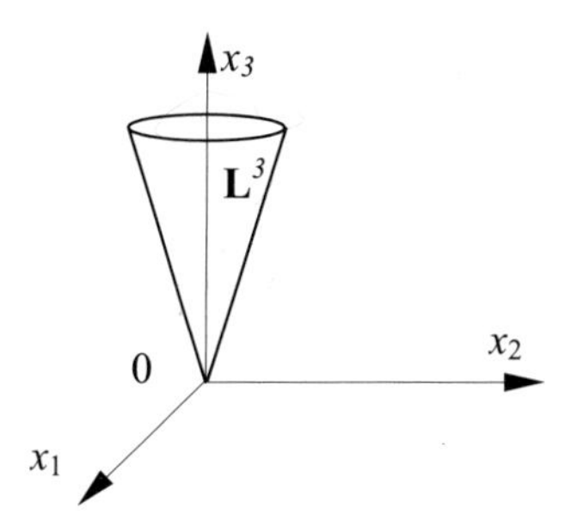

Figure 4.1: The Lorentz cone $\mathbf{L}^3$ is graphically shown in $\mathbf{R}^3$.

The positive semidefinite cone $\mathbf{S}^m_+$: this cone belongs to the space $\mathbf{S}^m$ of $m \times m$ symmetric matrices and is comprised of all $m \times m$ symmetric positive semidefinite matrices, i.e., of $m \times m$ matrices A such that

$$A = A^T, \qquad x^T A x \geq 0 \qquad \forall x \in \mathbf{R}^m.$$

4.3 Conic Programming

4.3.1 Conic programming under a primal form

Let **K** be a cone (convex, pointed, closed, and with a nonempty interior) in $\mathbf{R}^m$. Given an objective $c \in \mathbf{R}^m$, an $m \times n$ constraint matrix A and a right hand side vector $b \in \mathbf{R}^m$, the corresponding conic problem is the optimization program

$$c^T x \to \text{Min} \left| Ax - b \geq_{\mathbf{K}} 0 \right. \qquad \text{(CP)}$$

Note that the only difference between this problem and an LP program is that the latter deals with a particular choice of **K** with the case when **K** is the nonnegative orthant $\mathbf{R}^m_+$. Replacing this particular cone with other cones, we get the possibility to cover a lot of important applications, which cannot be captured by LP.

4.3.2 Conic Duality

We know from the duality in the LP that whenever a nonnegative weight vector λ satisfies the relation

$$A^T \lambda = c,$$

the inequality

$$\lambda^T Ax \geq \lambda^T b$$

yields a lower bound $b^T\lambda$ on the optimal value c^* in (LP). Also, the dual problem:

$$\text{Max } b^T \lambda \left| \lambda \geq 0, A^T \lambda = c \right.$$

was the problem of finding the best lower bound of the objective value in the LP.
The same scheme can be used to develop the dual to a conic problem:

$$c^T x \to \text{Min} \left| Ax \geq_{\mathbf{K}} b \right.$$

Here we should clarify what are the "admissible" weight vectors λ, i.e., the vectors such that the scalar inequality $\lambda^T Ax \geq \lambda^T b$ is indeed a consequence of the vector inequality $Ax \geq_{\mathbf{K}} b$. In the particular case of coordinate-wise partial ordering, i.e., in the case of $\mathbf{K} = \mathbf{R}^m_+$, the admissible vectors are those with nonnegative coordinates. Those vectors, however, are not necessarily admissible for ordering $\geq_{\mathbf{K}}$ given by cones **K** different from the nonnegative orthant.

For instance, consider the ordering $\geq_{\mathbf{L}^3}$ on $\mathbf{R}^3$ given by the three dimensional ice-cream cone:

$$\begin{pmatrix} a_1 \\ a_2 \\ a_3 \end{pmatrix} \geq_{\mathbf{L}^3} \begin{pmatrix} 0 \\ 0 \\ 0 \end{pmatrix} \Leftrightarrow a_3 \geq \sqrt{a_1^2 + a_2^2}\,.$$

Here, if vector a has its components as below, we have the inequality:

$$\begin{pmatrix}-1\\-1\\2\end{pmatrix} \geq_{\mathbf{L}^3} \begin{pmatrix}0\\0\\0\end{pmatrix} \quad (\Leftrightarrow \begin{pmatrix}-1\\-1\\2\end{pmatrix} \in \mathbf{L}^3, \text{ since: } 2 > \sqrt{(-1)^2 + (-1)^2} = \sqrt{2})$$

then by multiplying with a positive weight vector $\lambda = \begin{pmatrix}1\\1\\0.1\end{pmatrix}$ we will get a false inequality : $-1.8 \geq 0$

Thus not every nonnegative weight vector is admissible for the partial ordering $\geq_{\mathbf{L}^3}$.

We say that for a given cone **K**, the weight vector λ is admissable if

$$\forall a \geq_{\mathbf{K}} 0 : \lambda^T a \geq 0. \tag{4.1}$$

Whenever λ possesses the property (4.1), the scalar inequality

$$\lambda^T a \geq \lambda^T b$$

is a consequence of the vector inequality $a \geq_{\mathbf{K}} b$, which satisfies the axioms 1,...,4. Vice versa, if λ is an admissible weight vector for the partial ordering $\geq_{\mathbf{K}}$:

$$\forall (a, b : a \geq_{\mathbf{K}} b) : \lambda^T a \geq \lambda^T b,$$

then, of course, λ satisfies (4.1). Thus the admissable weight vectors λ are exactly the vectors from the set:

$$\mathbf{K}_* = \{\lambda \in \mathbf{R}^m : \lambda^T a \geq 0 \quad \forall a \in \mathbf{K}\}.$$

The set $\mathbf{K}_*$ is comprised of the vectors with nonnegative inner products with all vectors from **K** and called the dual cone to **K** (sometimes $\mathbf{K}_*$ is called the polar cone of the **K** cone). Geometrically, $\mathbf{K}_*$ and **K** are shown as below:

Figure 4.2: a. The angle between each vector λ in $\mathbf{K}_*$ and every vector a in **K** is less than or equal to $\pi/2$; b. $\mathbf{K}_* = \mathbf{K}$, **K** is a nonnegative orthant (figures are shown in $\mathbf{R}^2$).

After the theorem of the properties of the dual cone (see Ben-Tal and Nemirovski, 2001, p. 52), the set $\mathbf{K}_*$ has the following properties:

1. If **K** is a nonempty set, then $\mathbf{K}_*$ is a closed convex cone.
2. If **K** possesses a nonempty interior, then $\mathbf{K}_*$ is a pointed cone.
3. If **K** is a closed convex cone, then so is $\mathbf{K}_*$, and the cone dual to $\mathbf{K}_*$ is exactly **K**: $(\mathbf{K}_*)_* = \mathbf{K}$.
4. If **K** is a closed convex pointed cone, then $\mathbf{K}_*$ has a nonempty interior.
5. If **K** is a closed convex pointed cone with a nonempty interior, then so is $\mathbf{K}_*$.

Now, to derive the dual problem for a conic problem (CP) we define that whenever x is a feasible solution to (CP) and λ is an admissible weight vector, i.e., $\lambda \in \mathbf{K}_*$, x satisfies the scalar inequality:

$$\lambda^T Ax \geq \lambda^T b$$

and λ satisfies the relation:

$$A^T \lambda = c,$$

then one has:

$$c^T x = (A^T \lambda)^T x = \lambda^T Ax \geq \lambda^T b = b^T \lambda,$$

so that the quantity $b^T \lambda$ is a lower bound on the optimal value in (CP). The best bound one can get in such a manner is the optimal value in the problem:

$$\lambda^T b \to \text{Max} \left| A^T \lambda = c, \lambda \geq_{\mathbf{K}_*} 0 \right. \left[\Leftrightarrow \lambda \in \mathbf{K}_* \right] \tag{D}$$

and this program is the program dual to (CP).

Geometrically, we see that in (D) we are asked to maximize a linear objective $b^T \lambda$ over the intersection of the affine plane $\mathbf{L}_* = \{\lambda \mid A^T \lambda = c\}$ with the cone $\mathbf{K}_*$.

For the primal problem (CP): $c^T x \to \text{Min} \left| Ax - b \geq_{\mathbf{K}} 0 \right.$, one can prove that it geometrically merely minimizes a linear form, $d^T y$, over the intersection of an affine plan $\mathbf{L} = Ax - b = y$ with the cone $\mathbf{K}$.

Indeed, c can be represented by $A^T d$ in an assumption that the objective $c^T x$ can be expressed via $y = Ax - b$:

$$c^T x = d^T (Ax - b) + \text{const.}$$

This is equivalent to

$$c \in \text{Im}\, A^T \Rightarrow \exists d : c^T x = d^T (Ax - b) + d^T b, \ \forall x. \tag{4.2}$$

Under the premise of (4.2), the primal problem (CP) can be posed equivalently in terms of y, namely, as the problem

$$d^T y \to \text{Min} \left| y \in \mathbf{L}, y \geq_{\mathbf{K}} 0, \right. \quad \mathbf{L} = \left\{ y = Ax - b \middle| x \in \mathbf{R}^n \right\}.$$

Also, the feasible planes $\mathbf{L}$, $\mathbf{L}_*$ are orthogonal to each other.

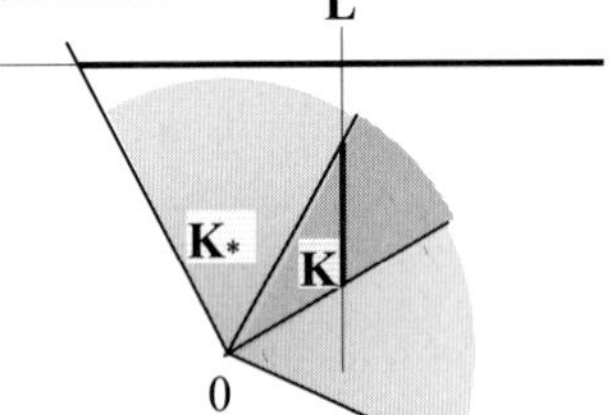

Figure 4.3: Primal dual pair of conic problems.

Summarizing, we have the conic primal problem (CP) and its conic dual problem (D) under the forms as:

$$c^T x \to \min \left| Ax - b \geq_{\mathbf{K}} 0 \right. \tag{CP}$$

$$\lambda^T b \to \max \left| A^T \lambda = c, \lambda \geq_{\mathbf{K}_*} 0 \right. \tag{D}$$

According to the conic duality theorem, we have the following properties:

1. The duality is symmetric: the dual problem is conic, and the problem dual to dual is primal.
2. The value of the dual objective at every dual feasible solution is less than or equal to the value of the primal objective at every primal feasible solution, so that the duality gap

   $$c^T x - b^T \lambda$$

   is nonnegative at every "primal-dual feasible pair" (x, λ), x is the primal feasible and λ is the dual feasible solution.
3. a. If the primal (CP) is below bounded and strictly feasible (i.e., $A^T x >_{\mathbf{K}} b$ for some x), then the dual (D) is solvable and the optimal values in the problems are equal to each other.

   b. If the dual (D) is above bounded and strictly feasible (i.e., $\lambda >_{\mathbf{K}_*} 0$ exists such that $A^T \lambda = c$), then the primal (CP) is solvable, and the optimal values in the problems are equal to each other.
4. Assume that at least one of the problems (CP), (D) is bounded and strictly feasible. Then a primal-dual feasible pair (x, λ) is comprised of optimal solutions to the respective problems

   a. if and only if $\quad b^T \lambda = c^T x \quad$ [zero duality gap]

   and

   b. if and only if $\quad \lambda^T [A^T x - b] = 0 \quad$ [complementary slackness]

4.4 Conic Quadratic Programming:

So far we have considered the conic programming problem in their generic form of the conic problems. Undoubtedly a most famous generic conic problem is the linear programming problem; however this is not the only interesting problem of this kind. The other fascinating generic conic problem of extreme importance is the Conic Quadratic program.

4.4.1 The primal-dual pair of conic quadratic problem

We have already seen that the m-dimensional Lorentz (≡second order ≡ ice cream) cone $\mathbf{L}^m$ is the cone given by

$$\mathbf{L}^m = \left\{ x = (x_1, ..., x_{m-1}, x_m)^T \in \mathbf{R}^m : x_m \geq \sqrt{x_1^2 + ... + x_{m-1}^2} \right\} \tag{*}$$

Here $m \geq 1$; in the case of $m = 1$ the empty sum is defined as $\sqrt{\sum_{i=1}^{0} x_i^2} = 0$, so that $\mathbf{L}^1$ is simply the nonnegative ray on the axis.

A conic quadratic problem is a conic problem:

$$c^T x \to \text{Min} \ \left| Ax - b \geq_{\mathbf{K}} 0 \right.,$$

for which the cone $\mathbf{K}$ is a direct product of several ice-cream cones:

$$\mathbf{K} = \mathbf{L}^{m_1} \times \mathbf{L}^{m_1} \times ... \times \mathbf{L}^{m_k}$$

$$= \left\{ y = \begin{pmatrix} y[1] \\ y[2] \\ ... \\ y[k] \end{pmatrix} \middle| y[i] \in \mathbf{L}^{m_i}, i = 1,...,k \right\}. \tag{4.3}$$

In other words, a conic quadratic problem is an optimization problem with linear objective and finitely many "ice-cream constraints"

$$A_i x - b_i \geq_{\mathbf{L}^{m_i}} 0, \quad i = 1,...,k,$$

where
$$[A,b] = \begin{bmatrix} [A_1,b_1] \\ [A_2,b_2] \\ ... \\ [A_k,b_k] \end{bmatrix},$$

is the partition of the data matrix $[A, b]$ corresponding to the partition of y in (4.3). Thus, a conic quadratic program can be written down also as

$$c^T x \to \min \left| A_i x - b_i \geq_{\mathbf{L}^{m_i}} 0, \quad i = 1,...,k. \right. \tag{4.4}$$

We know that the partial order $y \geq_{\mathbf{L}^m} 0 \Leftrightarrow y \in \mathbf{L}^m$ is the ordering such that the last component of vector y is greater than or equal to the Euclidean norm $\|.\|$ of the subvector of y comprised of the first $m-1$ components of y. Consequently, the $\geq_{\mathbf{L}^{m_i}} 0$ inequality can be written down as:

$$\|D_i x - d_i\|_2 \leq p_i^T x - q_i,$$

where

$$[A_i, b_i] = \begin{bmatrix} D_i & d_i \\ p_i^T & q_i \end{bmatrix}$$

is the partition of the data matrix $[A_i, b_i]$ into the sub-matrix $[D_i, d_i]$ comprised of the first $m-1$ rows and the last row $[p_i^T, q_i]$. We conclude that a conic quadratic problem can be written down as the optimization program

$$c^T x \to \text{Min} \left| \|D_i x - d_i\|_2 \leq p_i^T x - q_i, \quad i = 1,...,k, \right. \tag{QP}$$

where D_i are the matrices of the same row dimension as x, d_i are vectors of the same dimension as the column dimensions of the matrices D_i, p_i are vectors of the same dimension as x and q_i are reals.

It can be proved that (*) is a cone and, moreover, a self-dual cone: $\mathbf{K}_* = \mathbf{K}$. Consequently, the problem dual to (QP) is

$$\sum_{i=1}^{k} b_i^T \lambda_i \to \text{Max} \left| \sum_{i=1}^{k} A_i^T \lambda_i = c, \lambda_i \geq_{\mathbf{L}^{m_i}} 0, \quad i = 1,\ldots,k \right.,$$

since $\lambda_i \geq_{\mathbf{L}^{m_i}} 0$ and λ_i can be written as $\lambda_i = \begin{pmatrix} \mu_i \\ \nu_i \end{pmatrix}$ with vector μ_i and scalar component ν_i, then the form of the dual problem becomes:

$$\sum_{i=1}^{k} \left[\mu_i^T d_i + \nu_i q_i \right] \to \text{Max} \left| \sum_{i=1}^{k} \left[D_i^T \mu_i + \nu_i p_i \right] = c, \|\mu_i\| \leq \nu_i, \quad i = 1,\ldots,k \right. \qquad \text{(QD)}$$

The design variables in the dual form are vectors μ_i of the same dimensions as the vector d_i, and the scalars ν_i.

4.4.2 Conditions for a quadratic cone problem

In short, the standard forms of a conic quadratic problem for its primal and dual as (QP) and (QD) that are obtained as above will be applied to our optimization problem with the following conditions:

1. Assumption A: there is no nonzero x which is orthogonal to all rows of all matrices D_i and to all vectors p_i, $i = 1,\ldots, k$.
2. Strict feasibility of (QP) means that there exist x^* such that all inequality constraints $\|D_i x - d_i\|_2 \leq p_i^T x - q_i$ of the problem are satisfied by x^* as strict inequalities.
3. Strict feasibility of (QD) means that there exists a feasible solution $\{\bar{\mu}_i, \bar{\nu}_i\}_{i=1}^{k}$ to the problem such that $\|\mu_i\|_2 < \nu_i$ for all $i = 1,\ldots, k$.

4.5 Some important points for recognizing the conic quadratic problems

Before applying conic quadratic programming to our problems we need to know what the recognizing criteria are and how our problem can be cast in that framework. Here I simply mention some important points that will later be used to observe the problems in terms of their conic quadratic (CQ)-representable functions and sets.

Normally, an initial form of an applied optimization model is:

$$\min\{f(x) | x \in X\},$$

The set $X = \bigcap_{i=1}^{m} X_i$, is the set of designs admissible for all m design restrictions of X_i we take into the consideration. The set X_i in many cases is given by

$$X_i = \{x \in \mathbf{R}^n \,|\, g_i(x) \leq 0\}$$

If the objective function is non-linear, we can always pass on from this problem to an equivalent objective-constraint optimization problem.

$$t \to \text{Min} \left| (x,t) \in \hat{X} = \{(x,t) | f(x) - t \leq 0\} \cap \{(x,t) | x \in X_1\} \cap \ldots \cap \right.$$

Note that our new objective is linear in the new design variables (x, t).

Here we define the CQ-representable functions as the functions that possess CQ-representable epigraphs (see Ben-Tal and Nemiroski, 2001, p. 87-92).

$$\mathrm{Epi}\{f\} = \{(x,t) \in \mathbf{R}^n \times \mathbf{R} \,|\, g(x) \le t\}.$$

CQ-representability of a function g automatically implies CQ-representability of all level sets-the sets of the form $\{x \,|\, g(x) \le \mathrm{const}\}$.

4.5.1 Elementary CQ-representable functions/sets

The functions mentioned here are mainly the constraints in the inequalities or equalities.

1. A constant function $g(x) \equiv a$:
   The epigraph of a constant $\{(x,t) \,|\, a \le t\}$ is given by the linear inequality $0 \le t - a$, which is a special case of a conic quadratic inequality.
2. A linear function $g(x) = a^T x$:
   The epigraph of a linear function is given by the linear inequality $a^T x \le t$, or $0 \le t - a^T x$, and we can write this as the conic quadratic inequality: $\|0\|_2 \le t - a^T x$.
3. The Euclidean norm $g(x) = \|x\|_2$
   Indeed the epigraph of g is given by the conic quadratic inequality $\|x\|_2 \le t$.
4. The squared Euclidean norm $g(x) = x^T x$
   We can use $t = \left(\frac{t+1}{2}\right)^2 - \left(\frac{t-1}{2}\right)^2$, so that

$$x^T x \le t \Leftrightarrow x^T x + \left(\frac{t-1}{2}\right)^2 \le \left(\frac{t+1}{2}\right)^2 \Leftrightarrow \left\| \begin{pmatrix} x \\ \frac{t-1}{2} \end{pmatrix} \right\|_2 \le \frac{t+1}{2}$$

4.5.2 Operations preserving CQ-representability (CQr) of sets

1. Intersection: If sets $X_i \subset \mathbf{R}^n$, $i = 1, \ldots, m$, are CQr, so is their intersection $X = \bigcap_{i=1}^m X_i$.
2. Direct product: If sets $X_i \subset \mathbf{R}^{n_i}$, $i = 1, \ldots, k$, are CQr, then so is their direct product $X_i \times \ldots \times X_k$.

4.5.3 Operations preserving CQ-representability of functions

1. Taking maximum: if functions $g_i(x)$, $i = 1, \ldots, m$, are CQr, then so is their maximum $g(x) = \max_{i=1,\ldots,m} g_i(x)$.
2. Summation with nonnegative weights: if functions $g_i(x)$, $x \in \mathbf{R}^n$, are CQr, $i = 1, \ldots, m$, and α_i are nonnegative weights, then the function $g(x) = \sum_{i=1}^{m} \alpha_i g_i(x)$ also is CQr.

4.6 Applying SeDuMi 103 MATLAB toolbox for optimization over symmetric cones

4.6.1 A theoretical review of SeDuMi and its self-dual embedding technique

SeDuMi has been developed by Jos F. Sturm since 1998. This is an add-on for MATLAB, which enables optimization problems with linear, quadratic and semidefinite constraints to be solved. Large-scale optimization problems can be solved efficiently by exploiting the sparsity. SeDuMi stands for **Se**lf-**Du**al-**Mi**nimization. This software tool implements the famous interior point method that was initiated by Karmarkar in 1984. The interior point technique has been an important research area and used in optimization problems i.e., linear programming, quadratic programming, semidefinite programming, conic quadratic programming (second-order cone programming), convex non-linear programming and stochastic optimization.

We recall the standard form of the cone linear programming problems as:

$$\mathrm{Min}\{c^T x \mid Ax = b, x \in \mathbf{K}\} \tag{4.5}$$

Where $x \in \mathbf{R}^n$ is the vector of decision variables, $b \in \mathbf{R}^m$, $c \in \mathbf{R}^n$, $A \in \mathbf{R}^{m \times n}$ and $\mathbf{K} \subseteq \mathbf{R}^n$ is a specified convex cone.

And a dual problem:

$$\mathrm{Max}\ \{b^T y \mid A^T y + z = c, z \in \mathbf{K}^*\} \tag{4.6}$$

Where $y \in \mathbf{R}^m$ and $z \in \mathbf{R}^n$ are the vectors of decision variables, and

$$\mathbf{K}^* = \{ z \in \mathbf{R}^n \mid z^T x \geq 0 \text{ for all } x \in \mathbf{K} \}$$

is the dual cone to **K**.

The **K** cone can represent the nonnegative orthant ($\mathbf{K}=\mathbf{R}^n_+$), the cone of positive semidefinite matrices ($\mathbf{K}=\mathbf{S}^m_+=\mathbf{K}^s$), a Cartesian product of Lorentz cones ($\mathbf{K} = \mathbf{K}^q = \mathbf{L}^m_1 \times \mathbf{L}^m_2 \times \ldots \times \mathbf{L}^m_k = \mathbf{K}^q_1 \times \mathbf{K}^q_2 \times \ldots \times \mathbf{K}^q_{k(q)}$) or a symmetric cone (homogenous and self-dual cone), respectively. (see Sturm, 2002; Güler and Tunçel, 1998).

In SeDuMi, the Lorentz cone in $\mathbf{R}^n$ is defined as:

$$\mathbf{K}^q_i = \left\{ x \in \mathbf{R}^n \;\middle|\; x_1 \geq \sqrt{\sum_{i=2}^{n} x_i^2} \right\} \tag{4.7}$$

Generally, the interior point method used in the above problems can be summarized as follows (see Hillier and Lieberman, 1995):

1. Shoot through the interior of the feasible region toward an optimal solution.
2. Move in a direction that improves the objective value at the fastest possible rate.
3. Transform the feasible region to place the current trial solution near its center, thereby enabling a large improvement when implementing step 2.

The algorithm of the interior point method in SeDuMi is as follows:

Step 0. Initial solution $(x, y, z) \in \mathbf{K} \times \mathbf{R}^m \times \mathbf{K}$ with $Ax = b$ and $A^T y + z = c$ such that $\lambda(P(x)^{1/2} z) \in \mathbf{N}$, where λ is the spectral values of $P(x)^{1/2} z$ and $P(x)$ is the quadratic representative of x (see Sturm, 2002)

Step 1. If $x^T z \leq \varepsilon$ then STOP

Step 2. Choose Π being an invertible $n \times n$ block diagonal matrix and a vector $r \in \mathbf{R}^n$ according to the algorithmic settings (see Sturm, 2002). Compute the search direction $(\Delta x, \Delta y, \Delta z)$ from (a), (b) and (c).

$$\Delta x + \Pi \Delta z = r \quad , \tag{a}$$

$$A \Delta x = 0, \tag{b}$$

$$A^T \Delta y + \Delta z = 0. \tag{c}$$

Then determine a large enough step length $t > 0$ such that $\lambda\left(P(x + t\Delta x)^{1/2}(z + t\Delta z)\right) \in \mathbf{N}$.

Step 3. Update

$$(x, y, z) \leftarrow (x + t\Delta x, y + t\Delta y, z + t\Delta z)$$

and return to Step 1.

For step 1, one normally suggests an arbitrarily chosen vector in the interior of cone K as an initial point or the so-called "cold start" which means that no initial starting point is known (see Sturm, 2002). This initial point may or may not satisfy the linear feasibility constraints. The interior point method should then generate either an approximate primal-dual optimal solution pair, or an approximate Farkas-type dual solution to certify that no feasible solution pair exists.

The optimality conditions for (4.5) and (4.6) are:

$$b - Ax = 0 \tag{4.8}$$

$$A^T y + z - c = 0 \tag{4.9}$$

$$c^T x - b^T y \leq 0 \tag{4.10}$$

and

$$x \in \mathbf{K}, y \in \mathbf{R}^m \text{ and } z \in \mathbf{K}^* \tag{4.11}$$

The Farkas-type conditions to certify that there cannot exist (x, y, z) satisfying (4.8), (4.9) and (4.10) jointly are:

$$Ax = 0 \tag{4.12}$$

$$A^T y + z = 0 \tag{4.13}$$

$$c^T x - b^T y + 1 = 0 \tag{4.14}$$

together with (4.11).

The cold start interior point method is initialized from a triple ($x^{(0)}$, $y^{(0)}$, $z^{(0)}$) satisfying $\lambda(P(x^{(0)})^{1/2}z^{(0)})\in \mathbf{N}$. One may set $x^{(0)}$, $z^{(0)} = I$ and $y^{(0)} = 0$. One also defines:

$$y_0^{(0)} = \frac{\nu(\mathbf{K})+1}{\nu(\mathbf{K})} \tag{4.15}$$

where $\nu(\mathbf{K})$ is the order of a symmetric cone $\mathbf{K}$. For instance, the order of the nonnegative real half line is $\nu(\mathbf{R}_+) = 1$, each Lorentz cone has order $\nu(\mathbf{K}^q_i) = 2$ and because the Cartesian product of symmetric cones $\mathbf{K}_1$ and $\mathbf{K}_2$ has its order $\nu(\mathbf{K}_1\times\mathbf{K}_2)=\nu(\mathbf{K}_1)+\nu(\mathbf{K}_2)$ then the order $\nu(\mathbf{K}^q) = \nu(\mathbf{K}^q_1 \times \mathbf{K}^q_2 \times\ldots\mathbf{K}^q_{k(q)}) = 2k_{(q)}$ and the positive semidefinite cone has the order $\nu(\mathbf{K}^s) = \nu(\mathbf{K}^s_1 \times \mathbf{K}^s_2\times\ldots\times \mathbf{K}^s_{k(s)}) = \sum_{i=1}^{k(s)} \nu_i(s)$.

Given ($x^{(0)}$, $y_0^{(0)}$, $y^{(0)}$, $z^{(0)}$) to (4.8) and (4.9), the initial primal and dual residuals are defined as:

$$r_p = \frac{1}{y_0^{(0)}}(b - Ax^{(0)}), \qquad r_d = \frac{1}{y_0^{(0)}}(A^T y^{(0)} + z^{(0)} - c). \tag{4.16}$$

In SeDuMi, Sturm has used the self-dual embedding technique that was developed by Ye, Tood and Mizuno (1994). The slack variable z_0 is added to (4.10), and initialized at

$$z_0^{(0)} = \frac{(x^{(0)})^T z^{(0)}}{\nu(\mathbf{K})}, \tag{4.17}$$

and from (4.10) one computes

$$r_g = \frac{c^T x^{(0)} - b^T y^{(0)} + z^{(0)}}{y_0^{(0)}}, \quad x^{(0)} = 1.$$

The primal and dual problem (4.5) and (4.6) are embedded into a self-dual optimisation problem:

$$\text{Min } \{y_0 \mid (x^{(0)}, x, y^{(0)}, y, z^{(0)}, z) \text{ satisfies (4.19), (4.20), (4.21)}\} \tag{4.18}$$

With decision variables $x^{(0)}, x$, $y^{(0)}$, y, $z^{(0)}$, z, and the constraints (4.8), (4.9), (4.10) and (4.11) are rewritten under the form of (4.19)-(4.21) as

$$\begin{bmatrix} 0 & -A & b \\ A^T & 0 & -c \\ -b^T & c^T & 0 \end{bmatrix} \cdot \begin{bmatrix} y \\ x \\ x_0 \end{bmatrix} + \begin{bmatrix} 0 \\ z \\ z_0 \end{bmatrix} = y_0 \begin{bmatrix} r_p \\ r_d \\ r_g \end{bmatrix}, \tag{4.19}$$

$$r_p^T y + r_d^T x + r_g x_0 = 1 \tag{4.20}$$

$$(x^{(0)}, x) \in \mathbf{R}_+ \times \mathbf{K}, \qquad (y^{(0)}, y) \in \mathbf{R}^{1+m}, (z^{(0)}, z) \in \mathbf{R}_+ \times \mathbf{K}^*. \tag{4.21}$$

Premultiplying both sides of (4.19) with $[y^T \; x^T \; x_0]$ yields the identity

$$x^T z + x_0 z_0 = y_0(r_p^T y + r_d^T x + r_g x_0) = y_0. \tag{4.22}$$

Note that ($x^{(0)}$,x , $y^{(0)}$, y, $z^{(0)}$, z) satisfies (4.19)-(4.21) therefore it can be used as an initial starting point to solve (4.18) using a feasible inferior point method.

If we expand (4.19) into (4.23) with a given solution (x, y, z):

$$\begin{cases} b - A\hat{x} = (y_0 / x_0) r_b, \\ A^T \hat{y} + \hat{z} - c = (y_0 / x_0) r_c, \\ c^T \hat{x} - b^T \hat{y} < (y_0 / x_0) r_g, \end{cases} \tag{4.23}$$

where $\hat{x} = x/x_0$, $\hat{y} = y/x_0$, $\hat{z} = z/x_0$ are defined as components of a normalized solution

$$(\hat{x}, \hat{y}, \hat{z}) = \frac{(x, y, z)}{x_0}. \tag{4.24}$$

Equation (4.24) is the approximate solution to (4.8)-(4.11). When y_0/x_0 tends to zero, the residual to (4.8)-(4.10) tends to zero as well.

However, this is not always possible, since the original problem pair (4.5) and (4.6) can be infeasible. Therefore, another normalized solution is also defined:

$$(\tilde{x}, \tilde{y}, \tilde{z}) = \frac{(x, y, z)}{z_0} \tag{4.25}$$

as an approximate solution to (4.11)-(4.14). If we rewrite (4.23) into (4.26):

$$\begin{cases} -A\tilde{x} = (y_0 r_b - x_0 b) / z_0 \ , \\ A^T \tilde{y} + \tilde{z} = (y_0 r_c + x_0 c) / z_0 \ , \\ c^T \tilde{x} - b^T \tilde{y} + 1 = y_0 r_g / z_0 \ , \end{cases} \tag{4.26}$$

then we can see (4.26) is a similar form of (4.12)-(4.14). If, after the final iterations of the interior point method, the residual of $(\tilde{x}, \tilde{y}, \tilde{z})$ with respect to (4.12)-(4.14) is smaller than the residual of $(\hat{x}, \hat{y}, \hat{z})$ with respect to (4.8)-(4.10), the original problem is reported as infeasible, providing $\tilde{x}$ and $\tilde{y}$ as a certificate.

During the interior point process, one can predict whether the original problem pair is infeasible based on the (x_0, z_0) component of the first order predictor direction. The number *feas* is defined as:

$$feas = \frac{\dot{x}_0(0)}{x_0} - \frac{\dot{z}_0(0)}{z_0},$$

where the first order predictor direction is defined as the first order derivative with respect to t (the large enough step length) $\dot{x}_0(0) = \dfrac{dx_0(0)}{dt}$ and $\dot{z}_0(0) = \dfrac{dz_0(0)}{dt}$.

One can show that if a complementary solution exists then $\dot{x}_0(0) / x_0 \to 0$ and $\dot{z}_0(0) / z_0 \to -1$, so that $feas \to 1$. Conversely, if the problem is strictly infeasible one can show that $feas \to -1$. For problems without a complementary solution which are not strictly infeasible, this indicator is less valuable.

4.6.2 The general application of SeDuMi for the SW intrusion management problems

Firstly, SeDuMi must be correctly installed in the computer (see the installation instruction of SeDuMi 1.03 in the file " ...\SeDuMi103\install.dos"). Since our problems (SW intrusion management problems) are mainly in the field of linear programming and conic quadratic programming, we will therefore introduce some important points regarding the first two programs of SeDuMi for our application purposes.

d. Linear programming:

SeDuMi allows us to formulate the linear programming (LP) problem in either the primal standard form as follows:

$$\begin{aligned} &\text{Minimize} && c^T x && \text{(PSF1)} \\ &\text{such that} && Ax = b && \\ &\text{and} && x_i \geq 0 \quad \text{for } i = 1, 2, \ldots, n. && \end{aligned}$$

or the dual standard form:

$$\begin{aligned} &\text{Maximize} && b^T y && \text{(DSF1)} \\ &\text{such that} && c_i - a_i^T y \geq 0 \quad \text{for } i = 1, 2, \ldots, n. && \end{aligned}$$

When our problems are reformulated in those forms we have to add the slack variables. As defined in SeDuMi the artificial variables need not to be used and therefore only slack variables are inserted in the inequality constraints of the original-form of the problem which will become either the primal standard form (PSF1) or the dual standard form (DSF1) in SeDuMi, respectively.

After we transform our problem form to the standard form or a so-called "augmented form" of SeDuMi with the new b_1, c_1 vectors, and the new A_1 matrix, we will enter them in MATLAB. Now we can solve our problem in (PSF1) or (DSF1) form by invoking the function **sedumi**. Note that MATLAB is case sensitive and it is therefore essential to write **sedumi** in lowercase. For example, for the (PSF1) problem we type

```
>>sedumi(A1,b1,c1)
```

and after pressing enter, this functional command will solve the problem and show the optimal value of $c^T x$ and $b^T y$ and the optimal solution of x.

If we want to know the results of y we can type

```
>>[x,y]=sedumi(A1,b1,c1)
```

or if we want to know more information of the problem solved we insert a third argument called **info** in to the command

```
>>[x,y,info]=sedumi(A1,b1,c1)
```

The information (**info**) will be prompted with a MATLAB structure including the six fields:

cpusec:

iter:

feasratio:

numerr:

pinf:

dinf:

where **cpusec** is the field for the solution time information, **iter** for the number of iterations, and **feasratio** is the field for the final value of the feasibility indicator. There is also a field **numerr** which is nonzero in case of numerical problems (1 means premature termination: result are inaccurate, 2 means failure), and two fields **pinf**, **dinf** for the detected feasibility status of the optimization problem. If **pinf** =1 then the (PSF1) is infeasible, and y is a Farkas dual solution.

e. The conic quadratic programming:

In SeDuMi, it is possible to impose the quadratic constraints, by restricting variables to a quadratic cone. Such a restriction will replace the nonnegativity restriction in linear programming. Thus instead of requiring $x \in \mathbf{R}^n_+$ as in (PSF1), we will now require $x \in \mathbf{K}$, which is a so-called symmetric cone (see Güler and Tunçel, 1998). A symmetric cone is a Cartesian product of a nonnegative orthant, quadratic cone and cones of positive semidefinite matrices. For the conic quadratic programming the primal standard form for such an optimization problem is as follows

$$\begin{aligned} \text{Minimize} \quad & c^T x \\ \text{such that} \quad & Ax = b \\ \text{and} \quad & x \in \mathbf{K} \end{aligned} \tag{PSF2}$$

and the dual standard form is

$$\begin{aligned} \text{Maximize} \quad & b^T y \\ \text{such that} \quad & c_i - A^T y \in \mathbf{K} \end{aligned} \tag{DSF2}$$

Please recall that in SeDuMi, Sturm defines a quadratic cone as

$$\mathbf{Q}_{\text{cone}} = \{(x_1, x_2) \in \mathbf{R} \times \mathbf{R}^{n\text{-}1} \mid x_1 \geq \| x_2 \|\} \tag{QC}$$

where $\|.\|$ denotes the Euclidean norm. The quadratic cone is also known as the second order cone or Lorentz cone.

For example if we want to solve the problem as follows:

$$\text{Min } \{(t_1 + t_2) \mid t_1 \geq p_0^T y + \| Py \|,\ t_2 \geq \sqrt{2 + \|y\|^2}\ \}, \tag{ex.1}$$

where P is a given $n \times m$ matrix and p_0 is a given $m \times 1$ vector, then the decision variables are the scalars t_1 and t_2 and the vector $y \in \mathbf{R}^m$. The above problem has two quadratic constraints i.e.

$$(t_1 - p_0^T y, Py) \in \mathbf{Q}_{\text{cone}} \text{ and } (\ t_2, \begin{bmatrix} \sqrt{2} \\ y \end{bmatrix}) \in \mathbf{Q}_{\text{cone}}. \tag{ex.2}$$

The constraints of problem (ex.1) are similar to the constraint form of the dual problem (DSF2). Therefore, the problem (ex.1) can now be modeled under the form (DSF2) by constructing the two quadratic constraints in (ex.2) so that the whole problem can be formulated under the dual standard form of SeDuMi as below:

$$\text{Max} \qquad [-1 \ -1 \ 0].\begin{bmatrix} t_1 \\ t_2 \\ y \end{bmatrix}, \tag{ex.3}$$

such that

$$\begin{bmatrix} 0 \\ 0 \end{bmatrix} - \begin{bmatrix} -1 & 0 & p_0^{\mathrm{T}} \\ 0 & 0 & P \end{bmatrix} \cdot \begin{pmatrix} t_1 \\ t_2 \\ y \end{pmatrix} \in Q_{cone} \text{ with K.q} = [1+n], \tag{ex.4}$$

where K.q is defined as the field of the K structure that lists the dimensions of the quadratic cones. (The q in K.q stands for "quadratic".) The K-structure will be used to tell SeDuMi that the components of $(c - A^T y)$ are not restricted to be nonnegative as they would be in linear programming (DSF1). Instead, the first K.q(1) entries are restricted to a quadratic cone with $1+n$ components, and

$$\begin{bmatrix} 0 \\ \sqrt{2} \\ 0 \end{bmatrix} - \begin{bmatrix} 0 & -1 & 0 \\ 0 & 0 & 0 \\ 0 & 0 & -eye(m) \end{bmatrix} \cdot \begin{pmatrix} t_1 \\ t_2 \\ y \end{pmatrix} \in Q_{cone} \text{ with K.q=}[2+m], \tag{ex.5}$$

where **eye**(m) is a function in Matlab for creating the $m \times m$ identity matrix; the last K.q(2) entries are restricted to another quadratic cone with $2+m$ components.

Finally from (ex.3), (ex.4) and (ex.5) we have the total vectors:

$$b = \begin{bmatrix} -1 \\ -1 \\ 0 \end{bmatrix}, \qquad c = \begin{bmatrix} 0 \\ 0 \\ 0 \\ \sqrt{2} \\ 0 \end{bmatrix}, \tag{ex.6}$$

the total matrix:

$$A^T = \begin{bmatrix} -1 & 0 & p_0^{\mathrm{T}} \\ 0 & 0 & P \\ 0 & -1 & 0 \\ 0 & 0 & 0 \\ 0 & 0 & -eye(m) \end{bmatrix}, \tag{ex.7}$$

and the total K.q=[$1+n$, $2+m$], (ex.8)

with the total variables = $\begin{bmatrix} t_1 \\ t_2 \\ y \end{bmatrix}$.

After evaluating the total matrix and vectors from (ex.6)-(ex.8) we can solve the problem by invoking the command in SeDuMi as

```
>>[x,y,info]=sedumi(At,b,c,K)
```

where K is the structure to tell SeDuMi the number of quadratic constraints which is provided with values of the field K.q= [1+n, 2+m]. K is the new input argument in **sedumi** command. Without the fourth input argument, K, SeDuMi would solve a linear programming problem of the form (PSF1) or (DSF1).

To check that the solution of the problem satisfies the quadratic cones (ex.2), it is suggested to verify the inequality in the quadratic cone definition (QC) directly. However, it is more convenient to use the function **eigK**, which is part of SeDuMi. This function returns the eigen-values of a vector with respect to a symmetric cone. A symmetric cone consists of those vectors which have non-negative eigen-values. Therefore to check the feasibility and optimality we type in Matlab prompt:

```
>>[eigK(x,K),eigK(c-At*y,K)]
```

and

```
>>x'*(c-At*y)
```

Note that for the symmetric cone **K**, it holds that $x^T z \geq 0$ for all $x \in$ **K** and $z \in$ **K**, where $z = c - A^T y \in$ **K** and $x^T(c - A^T y) = c^T x - (Ax)^T y = c^T x - b^T y \geq 0$ (the complementary slackness and the duality gap in the conic duality theorem). Therefore, x provides an optimality certificate for y just as in the case of linear programming (see more details in Sturm 1998-2001).

In SeDuMi there is the possibility to incorporate both the nonnegative constraints and the quadratic cone constraints by inserting the nonnegative constraints in the first rows of the A^T matrix and the c vector. It is required that the nonnegative constraints must be transformed into the dual standard form (DSF1) of linear programming in SeDuMi as below:

$$c_1 - a_1{}^T y \geq 0 \qquad \text{(ex.9)}$$

For example if we would like to impose the nonnegativity on the variable vector $\boldsymbol{y} \geq 0$ in the problem (ex.1), we rewrite it as:

$$0 - (-\boldsymbol{y}) \geq 0$$

in the Matlab prompt we type:

```
>>c1=[zeros(m,1)]

>>a1=[zeros(m,2),-eye(m)]
```

and

```
>>K.l=[m]
```

where **eye**(m) is a function in Matlab for creating the $m \times m$ identity matrix, **zeros**(m,1) is a function in Matlab for creating the vector of m zero components and **zeros**(m, 2) for creating the $m \times 2$ matrix of $2m$ zero components. The K.l field, which is defined as the number of nonnegative linear constraints, will be inserted in the program to tell SeDuMi that the linear constraints are in the first l rows of the A^T matrix and the c vector. (The l in K.l represents "linear").

In SeDuMi, by convention the nonnegative constraints are always in the first components of the A^T matrix and the c vector. Therefore, the cone **K** is comprised of $\mathbf{R}^m_+ \times \mathbf{Q}_{cone} \times \mathbf{Q}_{cone}$ in the example problem. So, if inserting c_1 and a_1 into (ex.6) and (ex.7), the c vector and the A^T matrix become:

```
>>At=[a1;At]
```

```
>>c=[c1;c]
```

To solve the new problem we invoke again the **sedumi** command with the updated input arguments, A^T, c, K whereas b is kept the same.

```
>>[x,y,info]=sedumi(At,b,c,K)
```

Chapter 5

Development of the methodology for the multi-objective saltwater intrusion management model

5.1 Introduction

Based on the literature review, we have learned so far about the development of the saltwater intrusion management problem. In this work an approach that can solve the multi-objective saltwater intrusion management problem based on the sharp salt/fresh interface model is introduced. This simulation model has been applied to the regional-scale management problems for the areas where saltwater has intruded to their aquifers.

Besides the saltwater volume which has been used as a criterion for the saltwater intrusion management problem (Essaid, 1998), the saltwater intrusion length is another important output parameter from the model. Here, the saltwater intrusion length will be treated as one of the objectives for the management model.

This is the first time in the literature that this new approach has been proposed, i.e. the linkage between the simulation model and optimization program, which is mainly based on the response matrix of the saltwater (SW) intrusion length with respect to stresses. A second factor for the linkage i.e., the hydraulic head response matrix, which is already mentioned in the literature review, is also used in this work. If those response matrices are assumed constant (it is true for the response head matrix in cases of confined aquifers) with respect to the stress perturbation, and all the objective functions and constraints are linear, the linear programming for each single objective problem of the so called "deterministic multi-objective problem" with the assumed-linear response of the SW intrusion length will be applied. In order to solve the multi-objective problem there have been several techniques that mostly treat the problem under linear programming. The first type of technique is for generating non-inferior solutions that allow the analysts to communicate the results of the tradeoffs among objectives to the decision-makers. The second type is of techniques that incorporate preferences from decision makers and the third is of techniques with more involvement of the multi-decision makers. As the scope of this study is more that of analyst than of decision-maker, this study will concentrate on the first and second methods. Thus this multi-objective problem will be formulated by applying either the minimum distance from the ideal solution method or the weighting method. This problem will be solved by the quadratic cone programming, so-called Second Order Cone Optimization (SOCO) (see Sturm, 1998-2001 and Ndambuki, 2001) and the non-inferiority of its optimal solution will be checked by the constraint method (see Cohon, 1978).

As discussed in Chapter 3 concerning the confined aquifer, while the freshwater head at a control point responds linearly to the pumping (extraction or injection) rates, the saltwater intrusion length is a non-linear function with respect to those stresses. Hence, this will incur the complexity when applying even the linear programming to such management problems, since the coefficients of the objective functions and constraints are not always constant. However, to solve that problem the sequential-linearization approach, that was once introduced by Ahlfeld and Mulligan (2000) under a different form and approach, will be introduced. In this work, this new approach will allow the optimal solution for the problem with the non-linear response of the SW intrusion length to be searched out while seeking a suitable perturbing rate.

Due to the uncertainty in determining the aquifer parameters in the sense of e.g., the hydro-geological data, the problem will then be treated as a stochastic multi-objective optimization problem. In this work it is assumed that the multiple scenario approach (a variant of the classical Monte Carlo approach) has already been discussed by Ndambuki, (2001), which is relatively robust regardless of which scenario is actually realized. However, such solutions require consideration of a large number of realizations, resulting in a huge cost in terms of computational resources. This signals a form of trade-off between robustness and computational cost. Ndambuki (2001) proposed a more promising approach of dealing with uncertainty in the sense of randomly input parameters to the quantity groundwater management problem (for the confined aquifer only) through SOCO methodology.

Therefore, according to my knowledge, this is the first time in the literature that this new approach will be introduced to the SW intrusion management problem, a kind of quality groundwater management problems. This will enable the combination of the quadratic cone optimization (QCO or SOCO) and the sequential-linearization approach for solving the stochastic multi-objective optimization for the SW intrusion management problem. It will also involve more complications in terms of recalculating the response matrices for the number of realizations of the uncertain parameter while searching the robust optimal solution in an iterative manner.

5.2 Deterministic problem with the assumed-linear response of the SW intrusion length

It is known that in the confined aquifer the response of head with respect to stresses is rectilinear, while the response of the intrusion length is nonlinear. However, in this first approach the simplified problem will be formulated, based on the assumption of the linear response of the SW intrusion length with respect to stresses. The management problem here is comprised of the two linear objectives, i.e. Z_1 minimizing the total cost of operation and Z_2 minimizing the saltwater intrusion length. Those criteria functions are subject to several linear constraint inequalities. The linear vector-optimization problem will then have the form:

$$\underset{x}{\text{Min}}\ [Z_1 = c^T x], \tag{5.1}$$

$$\underset{x}{\text{Min}}\ [Z_2 = \lambda^T x], \tag{5.2}$$

subject to

$$\sum_{j=1}^{N_{ew}+N_{iw}} A_{i,j} x_j \geq b_i\ , \qquad i = 1,\ldots,N_c \tag{5.3}$$

$$\sum_{i=1}^{N_{ew}+N_{iw}} \lambda_i x_i \leq d\ , \tag{5.4}$$

$$e^T x \geq W_D\ , \qquad e^T = (1_1,\ldots,1_{New}) \tag{5.5}$$

$$x_i \leq Wc_i\ , \qquad i = 1,\ldots,N_{iw}+N_{ew} \tag{5.6}$$

$$x_i \geq 0 \quad , \qquad i = 1,\ldots,N_{iw}+N_{ew} \tag{5.7}$$

where

x	vector of decision variables, standing for the pumping (extraction or injection) rates Q_i , $[L^3T^{-1}]$,
c	vector of unit costs in monetary unit (MU), $c^T = (c_1,c_2)$,
c_1	unit cost for water extraction per time unit, $[MU.L^{-3}T^{-1}]$,
c_2	unit cost for water injection per time unit, $[MU.L^{-3}T^{-1}]$,
$A_{i,j}$	the response matrix of hydraulic head due to pumping rate Q_j at control point i. $A_{ij} = \dfrac{\partial h_i}{\partial Q_j}$, $[L^{-2}T]$, (5.8)
h_i	response head at control point i, [L],
b_i	the constraints values at control point i, $b_i = (h_i^{\min} - h_i^o)$, [L],
h_i^{min}	minimum head allowed at the control point i, [L],
h_i^o	head at the control point i under non-pumping conditions,
λ_i	the response vector of the average interface toe of the saltwater intrusion length, L, due to pumping rate Q_i, $\lambda_i = \dfrac{\partial L}{\partial Q_i}$, $[L^{-2}\,T]$, (5.9)
L	the intrusion length, [L],
d	the constraint values of the interface toe location, $d = (L^{\max} - L^o)$, [L],
L^{max}	maximum intrusion length allowed in case of aquifer restoration, [L],
L^o	intrusion length under non-pumping conditions, [L],
N_{ew}	number of extraction wells,
N_{iw}	number of injection wells,
N_c	number of control points,
W_D	total water demand, $[L^3T^{-1}]$,
Wc_i	maximum pumping rate allowed in well i, $[L^3T^{-1}]$.

5.2.1 Single Objective Problems

It is possible to easily solve each of the two single-objective problems by linear programming. Each single objective problem will be rewritten in an augmented form such that the linear programming solver in SeDuMi can recognize the problem and correctly solve it.

For instance, the augmented form of the primal standard form (PSF) in SeDuMi for the single objective problem is:

$$\text{Min} \qquad c^T x,$$

$$\text{such that} \qquad Ax = b,$$

$$x_i \geq 0 \text{ for } i = 1, 2, \ldots, n \qquad \text{(PSF)}$$

and its dual standard form (DSF) is:

$$\text{Max} \quad b^T y,$$

$$\text{such that} \quad c_i - a_i^T y \geq 0 \text{ for } i = 1, 2, \ldots, n, \qquad \text{(DSF)}$$

where $a_i^T = A^T$.

It is acceptable if each single objective problem is transformed into either the PSF or the DSF.

5.2.2 Multi-Objective Problem

For solving this multi-objective problem, the methods of minimum distance, goal programming and prior assessments of weights that incorporate the preferences will be considered. These methods can be combined with the constraint method, a type of generating technique that enables the analyst and decision-makers to numerically check the non-inferiority of the solution after all.

A. Methods for generating the non-inferior set

The constraint method

In multi-objective programming, non-inferiority is an important term for defining a state of the problem in which one can get a Pareto-optimal solution. The Pareto-optimal solution is defined as a solution where one cannot improve one of the objectives without deteriorating the other. Generally, in a multi-objective problem we can obtain many Pareto-optimal solutions upon which a so-called non-inferior set is defined. In other words, the non-inferior set is a set that is comprised of only the Pareto-optimal solutions. In linear programming, the non-inferior set is always located on a certain part of the boundary of the feasible region. In the minimizing multi-(two-) objective problem, the non-inferior set ($\mathbf{N}_o$) is found to the southwest of the feasible region in the objective space. This is called the "southwest rule". It is known that both of the feasible region and the non-inferior set can be shown either in the decision space as $\mathbf{F}_d$, $\mathbf{N}_d$ or in the objective space as $\mathbf{F}_o$, $\mathbf{N}_o$. Nevertheless, the south-west rule is *only* applicable in the objective space.

There is a number of methods in the techniques for generating the non-inferior set, e.g., the weighting method, the constraint method, the non-inferior set estimation method, etc. Here the constraint method for generating the non-inferior set is going to be used.

The mathematical background of the constraint method for a given multi-objective problem reads

$$\text{Min} \quad \boldsymbol{Z}(x_1, x_2, \ldots, x_n) = [\, Z_1(x_1, x_2, \ldots, x_n), Z_2(x_1, x_2, \ldots, x_n)], \qquad (5.10)$$

$$\text{s.t.} \quad (x_1, x_2, \ldots, x_n) \in \mathbf{F}_d\,. \qquad (5.11)$$

The constraint problem is

$$\text{Min} \quad Z_h(x_1, x_2, \ldots, x_n), \qquad (5.12)$$

$$\text{s.t.} \quad (x_1, x_2, \ldots, x_n) \in \mathbf{F}_d\,,$$

$$\text{and} Z_k(x_1, x_2,\ldots, x_n) \le L_k, \tag{5.13}$$

$$k = 1,\ldots, h-1, h+1,\ldots, p.$$

where

$$L_k = n_k + [t/(r-1)](M_k - n_k)\ , t = 0, 1, 2, \ldots, (r-1), \tag{5.14}$$

n_k, M_k are respectively the minimum and maximum values of the Z_k in the payoff table, so $n_k \le L_k \le M_k$.

The constraint problem of (5.11), (5.12), (5.13) is solved for every combination of values for the L_k, $k = 1,\ldots, h-1, h+1,\ldots, p$. Since r values of each of the objectives (except objective h which is in the objective function) will be used for calculating L_k, there are r^{p-1} combinations of the L_k. Each of r^{p-1} constraint problems requires the solution of a linear program and, if it is feasible, it will yield a non-inferior solution (if all objective constraints are binding). These solutions are the desired approximation of the non-inferior set (see Cohon, 1978).

For our multi-objective problem with two objectives, the objective Z_1 is chosen as an objective constraint and the problem of minimizing a single objective Z_2 will remain.

$$\min_x\ [Z_2 = \lambda^T x], \tag{5.2}$$

$$\text{s.t.} \quad x \in \mathbf{F}_d\ , \tag{5.11'}$$

$$Z_1 = c^T x \le L_k, \tag{5.15}$$

The least-squared algorithm

In the multi-objective problem, the least-squared algorithm is formulated in a sense of trying to find an optimal solution among the non-inferior set with the desirability of achieving more of all the objectives. However this algorithm can only be done after the non-inferior set is already generated. One can say this is a next step of the constraint method, even though it cannot be written in a standard form of an optimization problem.

In fact, this algorithm is based on the idea of the least-squared method. Here, instead of using the sum of squares of distances, δ is used as a square root of sum of squares of distances of $Z_j(x)$ from the end points Z^* of the non-inferior set and the least value of δ is also found by searching its minimum when varying $Z(x)$. This can be formulated as:

$$\text{Min}\ \delta = \min_{j=1,2,\ldots,r} \sqrt{\sum_{i=1}^{2} \left\| Z_i^* - Z_j(x) \right\|^2}\ , \tag{5.16}$$

s.t.

$$\forall Z^*,\ Z(x) \in N_o, \tag{5.17}$$

which are equivalent to

$$\text{Min}\ (\ \delta_j\ ), \tag{5.18}$$

s.t.

$$\sqrt{\sum_{i=1}^{n}\left\|Z_i^* - Z_j(x)\right\|^2} \le \delta_j , \qquad j = 1, 2, ..., r \tag{5.19}$$

$$\forall Z_i^*, \ Z_j(x) \in N_o$$

where $Z_j(x) = ([Z_1, Z_2, ..., Z_n]_j)^T$, n is the number of objectives and hence also the number of end points, Z_i^*, of the non-inferior set; r is the number of assumed extreme points in the non-inferior set generated by the constraint method. When $n = 2$ we have two end points, $Z_1^* = (Z_1^{opt}, Z_2^{trad})$, $Z_2^* = (Z_1^{trad}, Z_2^{opt})$ in which Z_1^{opt}, Z_2^{opt} are the optimal values of Z_1, Z_2 and Z_1^{trad}, Z_2^{trad} are the tradeoffs values of Z_1, Z_2 when the other objective is optimized. These points are calculated in the payoff table. $\mathbf{N}_o$ is the non-inferior set in the objective space. One can see that, in these formula forms, the linear constraints (3), (4), (5), (6), or (7) need not be reused.

The least-squared value, Min(δ_j), is here considered as also a criterion for searching the value of r in the constraint method in order to generate the non-inferior set as exactly as possible. This step is essential to the analysts and decision-makers, who want to check the solution or know the information of the non-inferior set before articulating their goals.

It is known that for the two-objective problem, the non-inferior set generated by the constraint method still much depends on the size of the number $(r - 2)$ being the number of extreme points in the approximate non-inferior set excluding the two-end points. This is because the constraint method with a certain small value of r will generate non-inferior solutions that might not be real non-inferior extreme points (see Cohon, 1978), even though the solution definitely lies on the real non-inferior set. So the question is then how large the r-value should be. To answer this question, the least-squared algorithm is introduced into the problem. It aims at making the non-inferior set finer by generating the maximum possible approximate extreme points with respect to the reasonably big size of r. Then a point, (Z_1(), Z_2()) where its least-squared value is found at such a non-inferior set, will be considered as a final optimal solution. This also means that, given a number of different sizes of r, we will obtain a corresponding number of r-dependent least-squared solutions and the final optimal solution will be achieved where the minimum value of those r-depended least-squared solutions is found. This approach will consume more computer time for searching this final optimal solution, otherwise the exact solution by the least-squared algorithm might not be achieved by the constraint method.

The procedure is as follows:

- Given a value of r, the constraint method will generate $(r - 2)$ extreme points (Z_1, Z_2) of the non-inferior set.

- By the least-squared algorithm (5.17), (5.18) and (5.19), one will find a smallest value (least square) $\delta^{min}{}_r$ = Min (δ_j) out of among δ_j each of which is the squared-root of the sum of squares of distances of the two-end points Z_1^* = (Z_1^{opt}, Z_2^{trad}), $Z_2^* = (Z_1^{trad}, Z_2^{opt})$ from the extreme point j. The δ^{min} indicates in the non-inferior set the point (Z_1, Z_2) at which, in this sense, its best compromised solution is found.

- Therefore, if given a series of different values of r, there will be a series of different values of $\delta^{min}{}_r$, and finally, the smallest value, δ^{final}, among the series of $\delta^{min}{}_r$ values will be found.

Thus far, this algorithm includes two steps:

- The first step is to calculate the least-squared value $\delta^{min}{}_r$ for a certain value of r in the series (3,..., k) (k is a big value), then one will find (k –2) values of least-squares, $\delta^{min}{}_r = [\delta^{min}{}_3, \delta^{min}{}_4, \ldots, \delta^{min}{}_k]$.
- The second step is to find among $\delta^{min}{}_r$ the minimum value δ^{final} from which the value of r and the least-squared solution are recorded correspondingly. This value of r is sufficient to generate the exact non-inferior set for checking the problem and the least-squared solution will be the final optimal solution in this sense.

In fact, in our application, for example, if a series of r-values vary from 3 to 100, then the δ^{final} becomes a constant (a fixed value) when r-values are around 54 upward (see Figure 5.3).

For instance, if $k = 10$ is chosen and r varies from 3 to 10 with step = 1, the minimum of least-squares is found at point $j = 3$ when $r = 4$, as illustrated in the Figure 5.1 and 5.2. With $r = 4$, the non-inferior set with two extreme points, as shown in Figure 5.2, is generated that might be different from the extreme points of the real non-inferior set.

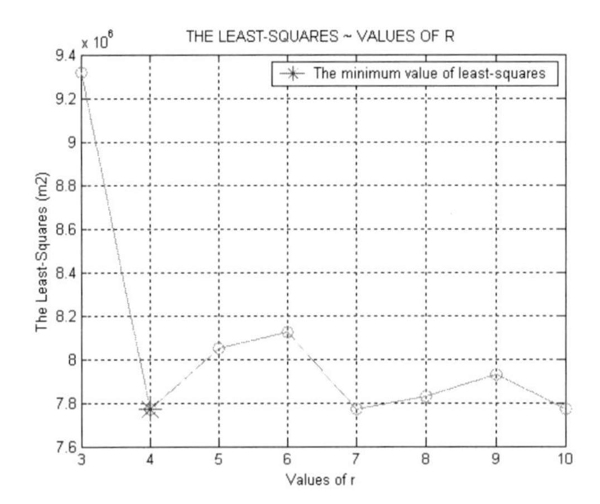

Figure 5.1: The minimum value of least-squares at $r = 4$ (r ranging from 3 to10).

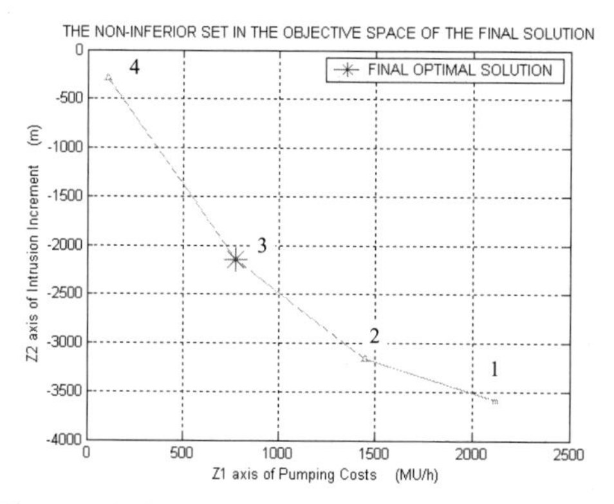

Figure 5.2: The non-inferior set with r = 4 and the least-squared optimal solution found at point 3.

However, if the series of r that varies from 3 to 70 is extended, one will find a smaller value of the minimal least square at j = 37 when r = 54, shown in Figures 5.3 and 5.4. Consequently, as shown in Figure 5.4, there will be the non-inferior set which is nearly identical to the real non-inferior set in the problem.

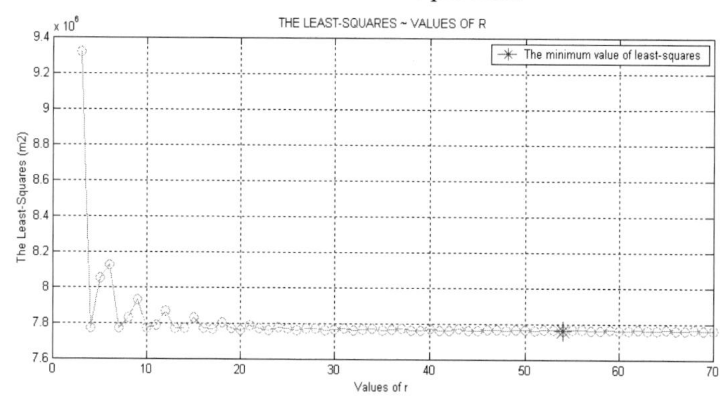

Figure 5.3: The minimum value of least-squares at r = 54 (r ranging from 3 to 70).

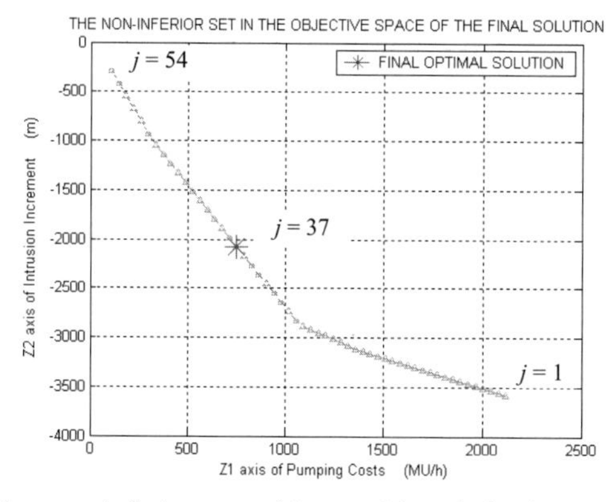

Figure 5.4: The non-inferior set with r = 54 and the least-squared optimal solution found at point j = 37.

This is true because if the range of r is extended up to 100 the same minimal least square that is also indicated at $r = 54$ will be found. This means that the real non-inferior set can be satisfactorily generated with $r = 54$ for the objective, which is to find the final solution for the two-objective problem by the least-squared algorithm.

B. Solution techniques that incorporate preferences

It is known that these solution techniques rely on explicit statements of preferences. These techniques include several methods in which the most usual methods e.g. the method based on the geometrical definitions of best and the method of prior assessments of weights, will be applied.

The methods based on the geometrical definitions of the notion of best

These methods are used for formulating the multi-objective problem and when the problem is solved, the best compromise will be achieved based on the geometrical meaning of the notion of best. This means the articulation will be stated in such a way that a feasible solution that is closest to the ideal solution on the basis of some distance measure is the desired solution. The distance mentioned here has only the geometrical meaning but no physical meaning at all. This is because the dimensions of objectives can be different from one to another. The methods of minimum distance from the ideal solution and the goal programming, which belong to these methods, will be mentioned here. In the minimum distance method, the metric has its general form as below:

$$||Z^*_k - Z_k||_\alpha = \left[\sum_{k=1}^{n} \left|Z_k^* - Z_k\right|^\alpha\right]^{1/\alpha} ;\alpha \in \{1,2,3,...\} \cup \{\infty\}, \qquad (5.20)$$

and the multi-objective problem can be rewritten as:

$$\text{Min } ||Z^*_k - Z_k||_\alpha , \qquad (5.21)$$

s.t.

$$x \in \mathbf{F}_\text{d} \qquad (5.11')$$

where:

$Z_k = [Z_1, Z_2,..., Z_k]^T$ is the vector objective;

$Z_k^* = [Z_1^*, Z_2^*,..., Z_k^*]^T$ is the ideal solution (utopia point), and $Z_1^*, Z_2^*,..., Z_k^*$ are the optimal values of objectives $Z_1, Z_2,..., Z_k$, respectively. Here $Z_1^*, Z_2^*,..., Z_k^*$ are the known constant-values.

If the multi-objective problem is minimizing Z_k then, for sure, $Z_k^* \le Z_k$. So it is possible to substitute the $| Z^*_k - Z_k |^\alpha$ by $(Z_k - Z^*_k)^\alpha$, and (5.21) becomes:

$$\text{Min } ||Z^*_k - Z_k||_\alpha = \text{Min}\left[\sum_{k=1}^{n} \left(Z_k - Z_k^*\right)^\alpha\right]^{1/\alpha}, \quad \alpha \in \{1,2,3,...\} \cup \{\infty\}, \qquad (5.22)$$

If $\alpha = 1$ then (5.22) becomes:

$$\text{Min}\left\|Z_k^* - Z_k\right\|_1 = \text{Min}\left[\left(Z_1 - Z_1^*\right) + ... + \left(Z_k - Z_k^*\right)\right], \qquad (5.23)$$

If $\alpha = 2$ then (5.22) becomes:

$$\text{Min}\left\|Z_k^* - Z_k\right\|_2 = \text{Min}\sqrt{\left(Z_1 - Z_1^*\right)^2 + ... + \left(Z_k - Z_k^*\right)^2}\,, \tag{5.24}$$

If $\alpha = \infty$ then (5.22) becomes:

$$\text{Min}\left\|Z_k^* - Z_k\right\|_\infty = \text{Min}\left[\text{Max}\left\{\left(Z_1 - Z_1^*\right), ..., \left(Z_k - Z_k^*\right)\right\}\right], \tag{5.25}$$

For the two-objective optimization problem in the deterministic case, its form with α = 2 is used:

$$\underset{x \in F_d}{\text{Min}}\left\|Z^* - Z(x)\right\|_2, \qquad \forall Z^*,\ Z(x) \in \mathbf{F}_o, \tag{5.26}$$

where $\mathbf{F}_d$ is the feasible region in the decision space that is created by the original constraints from (5.3),…,(5.7). $\mathbf{F}_o$ is the feasible region in the objective space. $Z^* = [Z_1^*, Z_2^*]^T$ together with its components, where Z_1^*, Z_2^* are the optimal solutions of the two corresponding objectives, Z_1, Z_2. The ideal solution (utopia point), (Z_1^*, Z_2^*) would be the solution of the multi-objective problem if there were no conflict among these objectives. $\| Z^* - Z(x) \|_2$ is the metric with α = 2. This metric measures the distance between the objective solution and the ideal point.

Ndambuki (2001) proposed $Z^* = [Z_1^*, Z_2^*]^T$ as the aspiration levels. Those terms are similar to the goals (target values) in the goal programming approach in which decision makers articulate the specification of goals. However, in goal programming the goals can be arbitrarily set corresponding to the decision makers' preferences if they do not know the non-inferior set beforehand. Consequently there may possibly be a set of goals that lead to an inferior solution. In goal programming, there can be one of the following four possible results:

- If the chosen goals are in the infeasible region to the southwest in the figure of the minimizing objective problem then the achieved solution will be non-inferior.

- Contrary to the above, the inferior solution can be obtained if the goals are also chosen in the infeasible region but to the northeast in Figure 5.5.

- If the chosen goals are the interior points of the feasible region in the objective space, those goals will be achieved because the minimum distance that is calculated by the goal programming is zero and hence the solution will be inferior.

- The zero distance is also obtained when the goals are in the non-inferior set. However in this case the solution will be non-inferior since those non-inferior goals are attained.

In the second and third cases, if the decision makers accept the obtained solution without checking the non-inferior set by the analyst, they will definitely settle for less than they should.

If there is communication between the analyst and decision makers or the analyst assumes the preferences for his multi-objective programming, the ideal solution that is always the infeasible solution is recommended to be proposed for our problem. This is because it guarantees that the best-compromise solution is always non-inferior by finding the minimum distance from the ideal solution. So the problem will be more oriented to the method of minimum distance from the ideal solution, with α = 2 in its metric. Please note that the objective being the metric with α = 2 is a non-linear function and as a result the multi-objective problem becomes a form of the optimization program with a non-linear objective.

For the two-objective problem, (5.26) is now reformulated under the non-linear-objective constraint problem:

$$\text{Min} \ \ \delta, \tag{5.27}$$

subject to

$$\left\| Z^* - Z(x) \right\|_2 \leq \delta, \tag{5.28}$$

$$\forall (x, \delta) \in \mathbf{F}_{\mathrm{d}}, \tag{5.29}$$

where $\mathbf{F}_{\mathrm{d}}$ is the feasible region in the decision space that is created by the linear constraints from (5.3),…,(5.7). $\left\| Z^* - Z(x) \right\|_2 \leq \delta$ is the non-linear-objective constraint. Now the new objective function is linearly determined on the set of design variable (δ, x).

As is known in the conic quadratic programming, there is a possibility of solving such a problem to find an optimal solution. This is when all of $Z_i(x)$ are linear functions, or the epigraphs of $g_i(x)$ that are given by the Euclidean norm can be rewritten in the conic quadratic inequality form (5.30) as below:

$$\text{Min} \ \delta,$$

subject to

$$\left\| q - Px \right\|_2 \leq \delta, \tag{5.30}$$

$$\forall (x, \delta) \in \mathbf{F}_{\mathrm{d}}.$$

Recall that (5.30) is in the form of (Qcone) a quadratic cone (a second order cone) or a Lorentz cone $\mathbf{L}^n$ by its definition, n is the number of its components. It is equivalent to the partial ordering on $\mathbf{R}^n$ of the vector $(\delta, q - Px) \geq_{\mathbf{L}^n} 0$ or (δ, $q - Px$)$\in \mathbf{L}^n$. The problem can now be rewritten under the dual standard form of the conic quadratic (CQ) program in SeDuMi:

$$\text{Max} \quad [-1, 0] \cdot \begin{bmatrix} \delta \\ x \end{bmatrix},$$

subject to

$$\begin{bmatrix} 0 \\ q \end{bmatrix} - \begin{bmatrix} -1 & 0 \\ 0 & P \end{bmatrix} \cdot \begin{bmatrix} \delta \\ x \end{bmatrix} \in Qcone$$

$$\text{and} \ \forall (x, \delta) \in \mathbf{F}_{\mathrm{d}},$$

where $q = \left[Z_1^*, Z_2^* \right]^T$, $P = \begin{bmatrix} c^T \\ \lambda^T \end{bmatrix}$ representing a two-objective problem and $\mathbf{F}_{\mathrm{d}}$ is the feasible region in the decision space that is created by the linear constraints from (5.3)

,..., (5.7). When $n > 2$ there is a general form of $q = \left[Z_1^*, Z_2^*, ..., Z_i^*, ..., Z_n^*\right]^T$ and $P = \left[c_1, c_2, ..., c_i, ..., c_n\right]^T$.

For example, in our application, the optimal values when $\alpha = 2$ are shown in Figure 5.5.

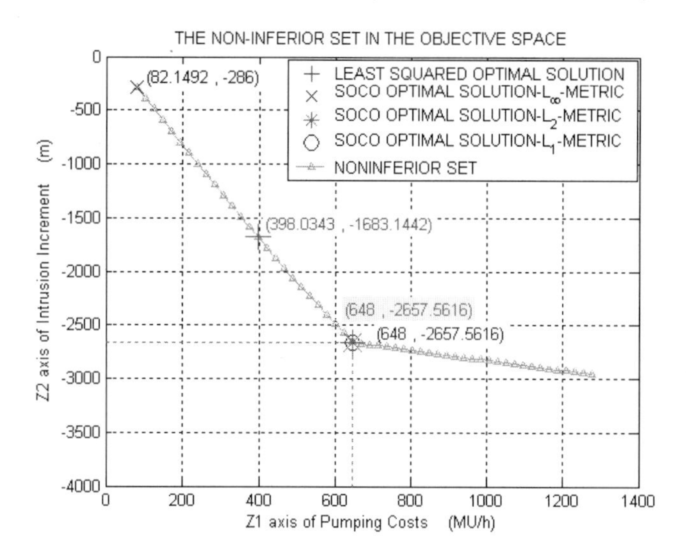

Figure 5.5: The optimal values of the two-objective problem with different L_α metrics: $\alpha = 1$, 2 and ∞ (The non-inferior is drawn for a fixed response vector of λ_j).

The method of prior assessments of weights

This method allows the decision-makers to articulate their preferences regardless of the non-inferior set being known or unknown. Similar to the weighting method in generating the non-inferior set, the prior assessments of weights on objectives is the usual way by which the preferences are stated. Differing from the goal programming, the weighting method will find the best compromise solution that is certainly the Pareto-optimal solution under some assumption. This means that the optimal solution obtained will surely be in the non-inferior set when the Kuhn-Tucker conditions for non-inferiority are necessarily satisfied and the weights w_k must be strictly positive for sufficiency in order to avoid the alternative optima where the optima can be inferior.

In the problem there are two objectives - the first objective is minimizing the operational costs of pumping and recharge, while the second objective is minimizing the saltwater (SW) intrusion length. Therefore, one can state that reducing the SW intrusion length for environmental improvement together with the protection of drinking water resource is w times more important than the money saved from the expenses for operating the recharge system. Then the multi-objective problem could be reduced to a single-objective problem and solved directly. The two-objective problem is formulated as:

$$\text{Minimize } \boldsymbol{Z}(w) = Z_1 + wZ_2 \tag{5.31}$$

s.t.

$$x \in \mathbf{F}_d, \tag{5.11'}$$

where w is a positive weight in (MU/m) unit.

In other words, stating that a distance decrement w_1 (in meters) of SW intrusion length is worth w_2 monetary unit (MU) is the same as the first statement. This latter statement can now be formulated as:

$$\text{Min} \quad Z(w_1, w_2) = Z_1 + (w_2/w_1)\, Z_2, \tag{5.32}$$

s.t.

$$x \in \mathbf{F}_d \tag{5.11'}$$

where w_1, w_2 are the positive weights in (m) and (MU) units, respectively.

Note that an objective function can be divided by a positive number without changing the problem, so (5.32) can be multiplied by w_1 to become

$$\text{Min} \quad Z(w_1, w_2) = w_1 Z_1 + w_2 Z_2. \tag{5.33}$$

Equation (5.33) is called the general form of the weighted problem where both objectives can be weighted.

Moreover, the ratio w_2/w_1 in (5.32) can be redefined as w and consequently (5.33) has the same form as (5.31). It must be kept in mind that the word ''worth'' here does not mean the real operational costs of recharging the aquifer for reducing the SW intrusion that one has to pay. Instead it means the willingness that the decision makers have to sacrifice an amount of the objective Z_1 for the objective Z_2.

Note that $(-1/w) = (-w_1/w_2)$ is defined as the marginal rate of substitution (MRS) or the desirable tradeoff that is the tradeoff between objectives of a utility function (an indifference curve). In other words, the amount of one objective that can be sacrificed for another objective is called the MRS (see Cohon, 1978). Besides that the feasible tradeoff is defined as the marginal rate of transformation (MRT) between Z_1 and Z_2, the negative slope of the non-inferior set ($\mathbf{N}_o$). The equality of the slope of $\mathbf{N}_o$ and the indifferent curve, MRT = MRS, is the tangency condition (Cohon, 1978).

At a certain level of $Z(w) = B$, the function in (5.31) can be rewritten as:

$$Z_2 = B/w + (-1/w) Z_1, \tag{5.34}$$

and clearly, $(-1/w)$ is the slope of the indifferent curve of a utility function (5.34). In (5.32) and (5.33) it can be seen that the weights are equivalent to a utility function that is linear in each objective and that expresses the constant MRS between objectives. The linearity implies that the marginal utility, which is a constant equal to the weight, does not decrease with the level of an objective. The constant MRS means that the willingness to tradeoff one objective for another is independent of the level of objectives.

Nevertheless, in linear programming only the extreme points among the non-inferior set are taken as the optimal solutions to the weighted problem (see Cohon, 1978). However, the method of prior assessments of weights gives more "safeness" to the decision makers than the method of goal programming in sense of the non-inferiority assurance or even more tradeoff possibilities (see Figure 5.6) than the method of minimum distance from the ideal point where the decision makers have only a narrow range (see Figure 5.5). This is in the so-called "compromise set" where preferences are articulated. Therefore, the possibilities for the decision makers stating their preferences in this method are to set the value of w. When w is assigned the multi-objective problem from (5.1) to (5.7) can be formulated as:

$$\text{Min} \qquad c^T x + w\lambda^T x \,, \tag{5.35}$$

subject to

$$x \in \mathbf{F}_{\mathrm{d}} \,. \tag{5.11'}$$

where $\mathbf{F}_{\mathrm{d}}$ is the feasible region in the decision space. $\mathbf{F}_{\mathrm{d}}$ is determined by the constraints from (5.3) to (5.7); w is the value of weight, $w > 0$.

The expression (5.35) is a linear objective that can be rearranged as $(c^T + w\lambda^T)x$. The whole problem will be solved with the PSF of the linear programming in SeDuMi. The optimal $\boldsymbol{Z}(w)$ at point (Z_1, Z_2) when $w = 1.5$, are shown in Figure 5.6.

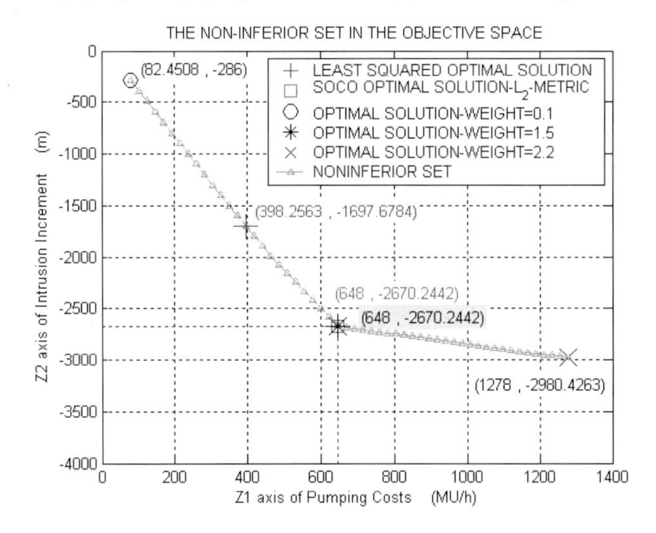

Figure 5.6: The optimal values of the two-objective problem with different weights: $w = 0.1$, 1.5 and 2.2 (The non-inferior is drawn for a fixed response vector of λ_j).

Finally, for the deterministic problem with the assumed-linear response of the intrusion length, a summary of the approach through several steps is as follows:

11. Setting up the quasi-three-dimensional model for the area, e.g. the geometric parameters, k, flux boundary, etc., as previously described.
12. Defining the deterministic management problem for the model area, e.g. the scenarios for the saltwater intrusion remedial management, as previously described.
13. Establishing the optimization problem with respect to the management scenarios, e.g. the objective functions, the constraints.
14. Running SHARP for that quasi-three-dimensional model with a fixed perturbed pumping rate (ΔQ) for every well location j to get the corresponding result file.

15. Based on the result files of SHARP, computing the response matrix of head $A_{ij} = [\Delta H_i / \Delta Q_j]$ and the vector of the averaged intrusion-length increment (the toe) $\lambda_j = [\Delta L / \Delta Q_j]$, the computing program PERTURB is written in FORTRAN.
16. Writing a program in Matlab to create the matrices, e.g. A, λ, b, c, etc., for the linear programming.
17. Running the linear programming in SeDuMi (written in Matlab) to solve the (two) problems with each single objective.
18. Using the constraint method to generate the non-inferior set based on the optimal solutions with respect to these objectives (here two objectives).
19. Either solving the multi-objective problem with the method of minimum distance from the ideal solution or the utopia point (the point is comprised of two optimal solutions of the two single-objective problems). If formulating the problem with the L_2-metric the conic quadratic programming (SOCO) in SeDuMi is used as a solver.
20. Or solving the multi-objective problem with the method of prior assessments of weights. The problem will be solved with the linear programming in SeDuMi.

In fact, the solutions of step 9 and 10 do not, however, need step 8. Step 8 is just for the numerical check of the non-inferiority of the solutions.

5.3 Deterministic problem with a non-linear response of the intrusion length

To incorporate SHARP with SeDuMi, the response matrixes of head and intrusion lengths will be created in order to link those two programs. For the confined aquifers the components of the head response matrix is really constant. However, because the relation between the stress and the intrusion length is non-linear (see fig. 1a, 1b), the components of the intrusion length response matrix are not constant. If using the Taylor series to relate the nonlinear response of the intrusion length to pumping rates, the first order is seen to be not exact.

$$L^{k+1} = L^k + \sum_{j}^{N_{ex}+N_{in}} \left[\frac{\partial L}{\partial Q_j} \cdot \left(Q_j^{k+1} - Q_j^k \right) \right] + error \tag{5.36}$$

where the L^{k+1} and L^k are the values of the intrusion length at the $(k+1)^{th}$ and k^{th} iterations.

Q^{k+1} and Q^k are the pumping rates at the $(k+1)^{th}$ and k^{th} iterations.

$$\frac{\partial L}{\partial Q_j} \approx \frac{\Delta L}{\Delta Q_j} = \frac{\left[L(Q + \Delta Q_j) - L(Q) \right]}{\Delta Q_j} \tag{5.37}$$

$\Delta L / \Delta Q_j$, that is called the response coefficient or the jth element of the response vector, λ_j, depends on the Q_j and ΔQ_j in the kth iteration. The response coefficient is likely to be the gradient of the functional relationship between the intrusion length and the pumping rates at a well.

The error is a sum of the partial derivatives of the second order and higher as below:

$$error = \sum_{i}^{N_{ex}+N_{in}} \sum_{j}^{N_{ex}+N_{in}} \left[\frac{\partial^2 L}{\partial Q_i \, \partial Q_j} \cdot \frac{\left(Q_i^{k+1} - Q_i^k \right)\left(Q_j^{k+1} - Q_j^k \right)}{2!} \right] + \ldots$$

For a small extension of stress or a simple case, one can assume that the $L\sim Q$ relationship is linear (*error* = 0 in Taylor series) and the first order is exact. It aims at constructing the response vector, $\Delta L / \Delta Q_j$, with its constant components, and consequently that is the assumption for the first problem that has been tackled.

In general, the sequential linearization approach should be introduced into the problem when the stress is increasingly applied such that the linear relationship no longer holds. In case of a small difference of $(Q_i^{k+1} - Q_i^k)$, the error term can be neglected because of its second order.

At $k = 1$, when given initial values of Q_j^1 and a small perturbation ΔQ_j then the initial L^1 and the response vector $\lambda_j^1 = (\partial L / \partial Q_j)^1$ will be computed. With these λ_j^1 gradients the optimization program will solve the linear problem for its optimal solution Q_j^2 and hence the objective value Z_2^2 by the objective (5.2):

$$Z_2^2 = \lambda_j^1 \{Q_j^2\}$$

Also from the optimal solution Q_j^2 the simulation model will update its SW intrusion length L^2. Note that for the kth iteration, the coefficients of the objective (5.2) and the constraint (5.4) in the linear program are exactly the components of the vector λ_j^k. These will be altered in the next $(k+1)^{th}$ iteration if the norm of difference between two consecutive values of (Q_j^k, Z_2^k, L^k) and (Q_j^{k+1}, Z_2^{k+1}, L^{k+1}) are greater than ε. However, the difference between Z_2^k and Z_2^{k+1} can only be evaluated from the second iteration onward since during the first iteration ($k = 1$) the optimal value Z_2 has just been computed first.

It can also be seen that the optimal value Z_2^{k+1} should be equal to L^{k+1} for every iteration.

$$L^{k+1} \approx Z_2^{k+1} = \lambda_j^k \{Q_j^{k+1}\}$$

Therefore $$L^{k+1} \approx \lambda_j^k \{Q_j^{k+1}\} \tag{5.38}$$

And if $\lambda_j^k \approx \lambda_j^{k-1}$ the following can be written:

$$L^k \approx \lambda_j^k \{Q_j^k\}$$

or $$L^k - \lambda_j^k \{Q_j^k\} \approx 0 \tag{5.39}$$

It is true if one sets $Q_j^1 = 0$ and $L^1 = 0$ at the first iteration ($k = 1$). Adding (5.39) to the right side of (5.38) produces:

$$L^{k+1} \approx \lambda_j^k \{Q_j^{k+1}\} + L^k - \lambda_j^k \{Q_j^k\}$$

or

$$L^{k+1} \approx \lambda_j^k \{Q_j^{k+1} - Q_j^k\} + L^k \tag{5.40}$$

Recall that (5.40) is the same form of the Taylor series for the first order (5. 36):

$$L^{k+1} \approx L^k + \sum_{j}^{N_{ex}+N_{in}} \left[\left(\frac{\partial L}{\partial Q_j} \right)^k \left(Q_j^{k+1} - Q_j^k \right) \right] \tag{5.36}$$

When the algorithm is converged, the values of L^{k+1} will be close to L^k, $Q_j^{k+1} \approx Q_j^k$ and $Z_2^{k+1} \approx Z_2^k$.

5.3.1 The sequential linearization approach

The linearization approach introduced here is based on the iterative procedure, called sequential linearization. For the first iteration an initial state of the aquifer interface without extraction or injection will be established. Then in turn a small constant perturbation, ΔQ_j, of the extraction/injection rate (with $Q_j = 0$) at each candidate well location will be applied. The SHARP and PERTURB will run to create the response matrices $A_{ij} = [\Delta H_i / \Delta Q_j]$ and $\lambda_j = [\Delta L / \Delta Q_j]$ (an initial gradient value at the first end point of the curve). These matrices are used for the first run of SeDuMi to find the first iteration's optimal solution, $\boldsymbol{Q}_{opt}^1$, for the multi-objective problem, Z (Z_1, Z_2). In fact, the matrix A_{ij} need not be repeatedly computed, since A_{ij} is a matrix with constant elements when the aquifer is confined.

The optimal solution, $\boldsymbol{Q}_{opt}^1$, brings about the values of the objective Z_1^1 (pumping costs) and Z_2^1 (the SW intrusion increment/decrement) that are the coordinates of the best-compromised point in the non-inferior set. Based on that optimal solution (the extraction/injection rates at all candidate wells) SHARP will run again to give the simulation result of the SW intrusion increment/decrement, L_{sim}^1, which will be then compared with the optimal intrusion increment value in the next iteration. The results (Z_2^1, $\boldsymbol{Q}_{opt}^1$, L_{sim}^1) at the end of the first iteration will be remembered.

By taking the values of $\boldsymbol{Q}_{opt}^1$ and the same ΔQ, the second iteration will start with the same procedure as the first iteration. In the second iteration the new vector λ_j (the gradient values at the second points of the curves) are newly computed with the stress $Q_j = \boldsymbol{Q}_{opt}^1$ and the same perturbation ΔQ. At the end of this iteration the new optimal solution, (Z_2^2, $\boldsymbol{Q}_{opt}^2$, L_{sim}^2), will be obtained and then compared with the solution of the first iteration. If the norm of the difference between two consecutive iterations is greater than an allowable range of the error, ε, the program will start over again with the next (3rd) iteration, an updated set of extraction/injection rates by taking the recent optimal solution, $\boldsymbol{Q}_{opt}^2$. Together, the perturbed extraction/injection rate ΔQ is also used to create the new response vector, λ_j (another gradient value at the next point of the curve). Otherwise that optimal solution will be the final optimal solution for the non-linear case of the SW intrusion length.

To be clear the whole procedure is summarized as follows:

1. Set $k = 1$, assume the initial salt/freshwater interface toe, L_{sim}^1 as the initial state of non-pumping condition ($\boldsymbol{Q}^1 = 0$).
21. Compute the response coefficients of the vector λ_j.
22. Solve the multi-objective program for the optimal solution $x = \boldsymbol{Q}_{opt}^{k+1}$

$$\text{Min} \qquad [Z_1 = c^T x\,], \tag{5.1}$$

$$\text{Min} \qquad [Z_2 = \lambda^T x], \tag{5.2}$$

such that

$$A_i x \geq b_i, \tag{5.3}$$

$$\lambda^T x \leq d, \tag{5.4}$$

$$e^T x \geq W_D, \tag{5.5}$$

$$x \leq Wc, \tag{5.6}$$

$$x \geq 0. \tag{5.7}$$

23. Use $\boldsymbol{Q}^{k+1} = \boldsymbol{Q}_{opt}{}^{k+1}$ to compute the values of the objectives $Z_1{}^{k+1}$ and $Z_2{}^{k+1}$.
24. Use $\boldsymbol{Q}^{k+1}$ to update the saltwater intrusion length, $L_{sim}{}^{k+1}$.
25. if $\left\| \left\{ \boldsymbol{Q}^{k+1}, Z_2{}^{k+1}, L_{sim}^{k+1} \right\}^T - \left\{ \boldsymbol{Q}^k, Z_2{}^k, L_{sim}^k \right\}^T \right\| \leq \varepsilon_1$, or $\left\| \left\{ \boldsymbol{Q}^{k+1} \right\}^T - \left\{ \boldsymbol{Q}^k \right\}^T \right\| \leq \varepsilon_2$, (in the last case attention is only paid to the convergence of the optimal solution), stop; otherwise go to step 7
26. Set $k = k+1$; go to step 2.

For this summary there is a flow chart in Figure 5.7 that shows the whole procedure of this approach.

5.3.2 The convergence of the sequential linearization approach

Graphically, the convergence of the problem will be considered, for example, by the observation of the two response curves of the intrusion length versus the extraction and injection rates. To enable a graphical observation, the optimization problem will be set with two variables, mainly, x_1 is the pump rate for the only extraction well (Q_1) and x_2 is the pump rate for the only injection well (Q_2). Then the total number of wells is $N_w = N_{ew} + N_{iw} = 1+1 = 2$. For simplification there is only one control point chosen for the head constraint (this control point should be selected landward at the cell next to the extraction well), then the number of control points is $N_c = 1$. So the example problem from (1) to (7) now has a specific form:

$$\text{Min} \quad [Z_1 = 0.01x_1 + 0.1x_2] \tag{5.1'}$$

$$\text{Min} \quad [Z_2 = \lambda_1 x_1 + \lambda_2 x_2] \tag{5.2'}$$

such that

$$(-0.0144)x_1 + (0.0011)x_2 \geq b = -5 \tag{5.3'}$$

$$\lambda_1 x_1 + \lambda_2 x_2 \leq d = -10 \leq 0 \tag{5.4'}$$

$$x_1 + 0\, x_2 \geq W_D = 50 \tag{5.5'}$$

$$x_1 \leq Wc_{\text{extraction}} = 200 \tag{5.6'}$$

$$x_2 \leq Wc_{\text{injection}} = 900 \tag{5.6''}$$

$$x_1, x_2 \geq 0. \tag{5.7'}$$

Here b is the allowable draw-down of head ($b = -5$ indicates that the lowest designed head is 5m lower than the head at non-pumping state) and d is the upper bound of the SW intrusion increment ($d = -10$m indicates that the maximum SW intrusion length is 10m shorter than the SW intrusion length at non-pumping condition). Hence, the constraint (5.4') means that the designed SW intrusion length must be at least 10 m shorter than the SW intrusion length at non-pumping condition. Water demand and well capacities are represented by W_D and Wc respectively.

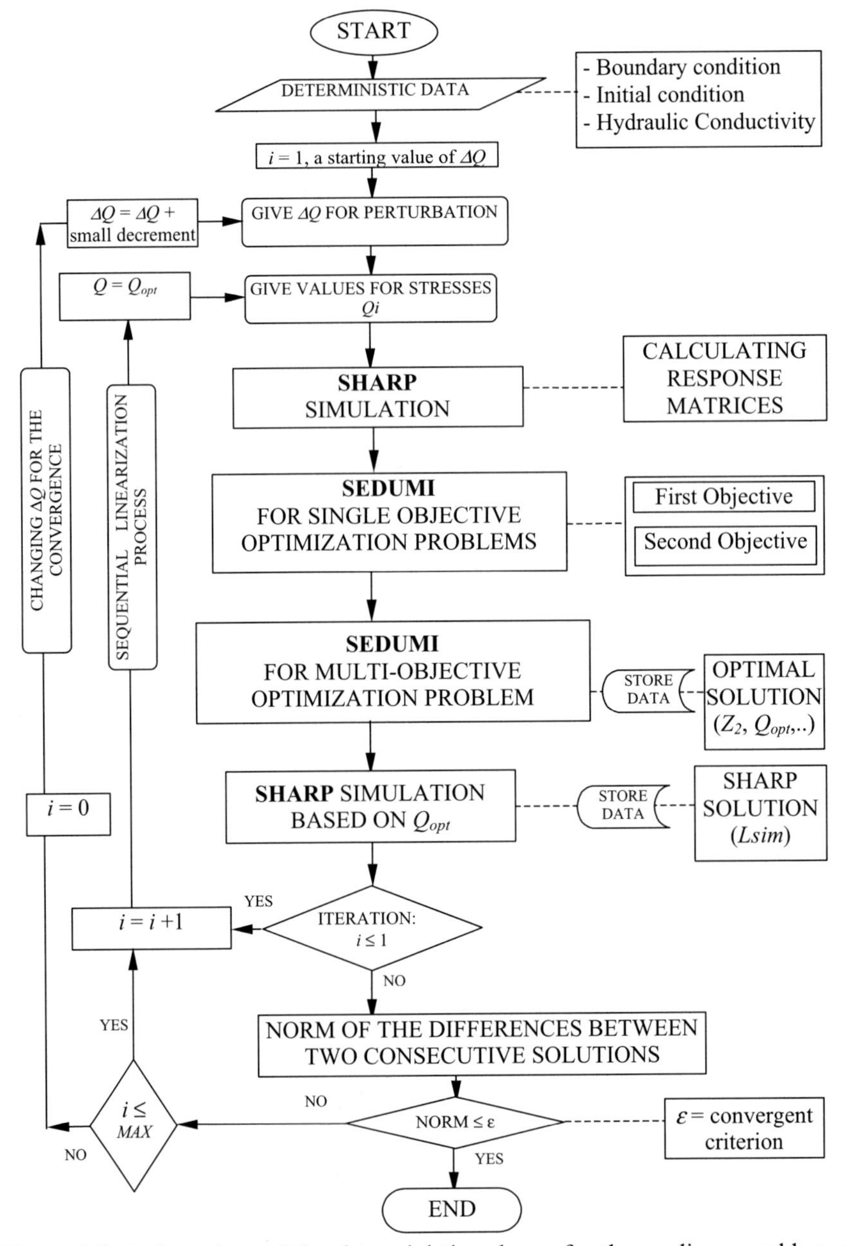

Figure 5.7: A flow chart of the deterministic scheme for the nonlinear problem of saltwater intrusion management

In the objective (5.2') and the constraint (5.4'), their coefficients are the components of the vector $\lambda^T = (\lambda_1, \lambda_2)$, $i = 1,\ldots, N_w = 2$. These vary differently depending on the iteration when Q_i varies, while the coefficients in the constraint (5.3') are the constant components of the row matrix $A_{1i} = (-0.0144, 0.0011)$.

The feasible region

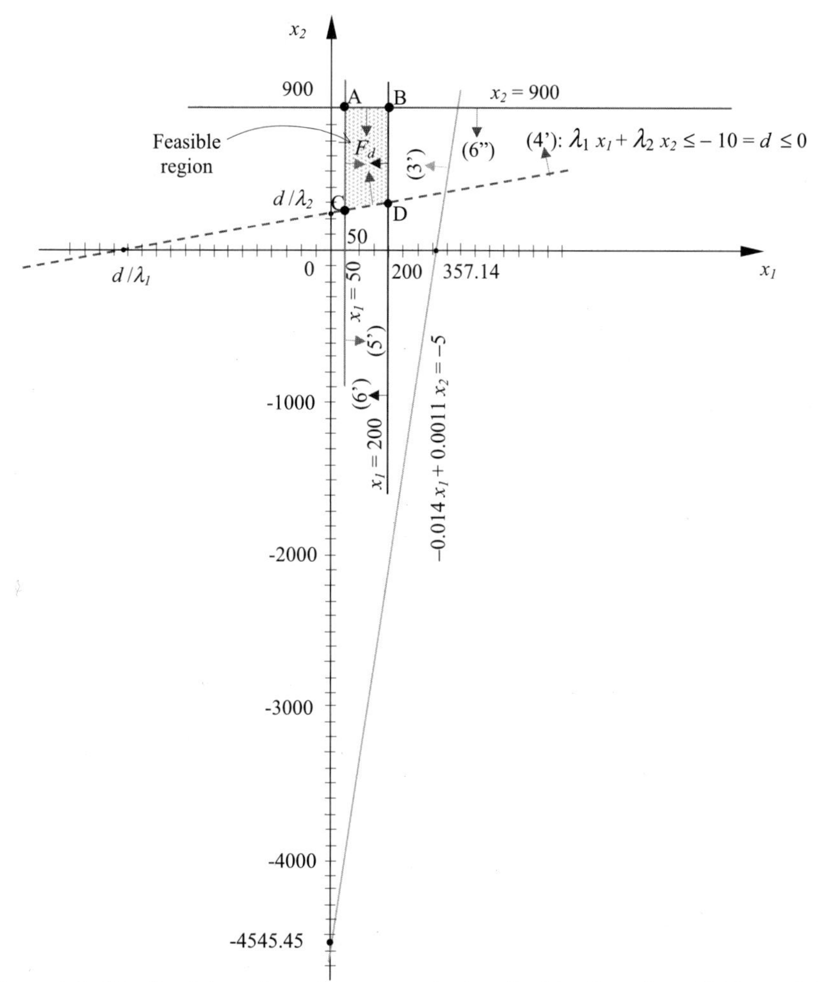

Figure 5.8: The feasible region of the example described by Equations (5.1'),…,(5.7')

Note that λ_1 is the response coefficient of the extraction well; therefore λ_1 is always set positive. Whereas λ_2 is the response coefficient of the injection well, hence, it is set always negative. Here $d = L_{max} - L_o$ where L_{max} is the maximal SW intrusion length allowable in the planning and L_o is the initial SW intrusion length at the non-pumping condition. Since L_{max} is set less than or equal to L_o in the case that more freshwater in the aquifer has to be stored, then d is less than or equal to zero. The constraint (5.4') is then understood that the SW intrusion length affected by extraction is always less than the expelled SW length induced by injection.

If the constraints (5.3'), (5.5'), (5.6'), (5.6'') and (5.7') are drawn in the decision space (x_1, x_2) one obtains the feasible region, $\mathbf{F}_d$, that is the trapezoid *ABCD* (the shaded area) which is the intersection of only four closed half-spaces (5.5'), (5.6'), (5.6'') and (5.7') shown in Figure 5.8. In this problem the constraint (closed half-space) (5.3') is a redundant constraint that does not matter whether it exists or not. In the feasible region, the extreme points *A* and *B* are fixed since they are the intersection points of the line (hyper plane) of $x_2 = 900$ in constraint (5.6'') with lines $x_1 = 50$ and $x_1 = 200$ in constraints (5.5') and (5.6') respectively. Whereas the extreme points *C* and *D*, which are the intersection points of the lines $x_1 = 50$ and $x_1 = 200$ with the line $-10 = \lambda_1 x_1 + \lambda_2 x_2$ in constraint (5.4'), will move dependently on the movement of (5.4') when the iteration progresses. In Figure 5.8, the value of $d / \lambda_1 \leq 0$ is always in the negative part of the x_1 axis and $d/\lambda_2 \geq 0$ is always in the positive part of the x_2 axis.

The second single objective (Z_2) problem

If the second objective function (Z_2) in the same figure of the feasible region in the decision space is drawn, the Figure 5.9 is produced:

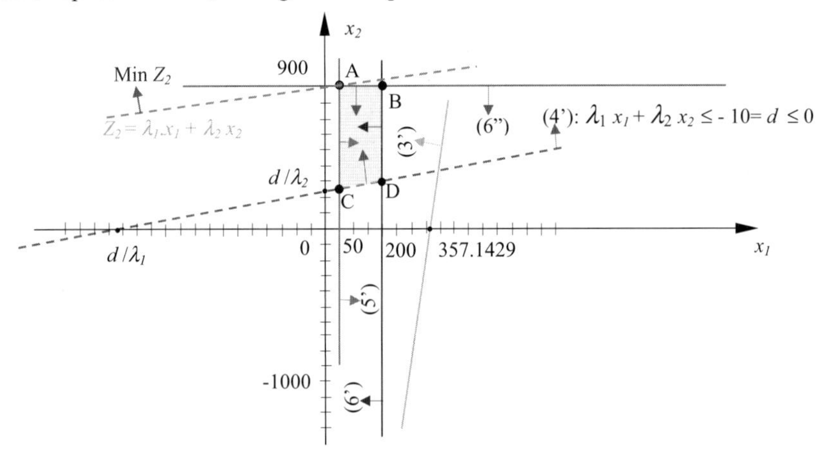

Figure 5.9: Graphical representation of the second objective function and its optimal solution at point *A*.

Since λ_1 is always positive and λ_2 is always negative regardless of iteration, the gradient of the contour lines of Z_2 is the vector (λ_1, λ_2) that always directs the arrow to the southeast of the (x_1, x_2) space. Therefore, the direction for minimizing Z_2 is to the northwest as shown in Figure 5.9. Consequently, the optimal solution for Z_2 is at the extreme point *A* which has its coordinates as ($x_1 = 50$, $x_2 = 900$). This solution is kept unchanged when the iteration progresses.

The first single objective (Z_1) problem

In Figure 5.10 it can be seen that the gradient of contour lines (hyper planes) of Z_1 has the components as (0.01, 0.1) whose arrow is headed to the northeast and hence the direction indicating the minimization of Z_1 is always to the southwest. In this problem the optimal solution will be at the extreme point, *C*, which depends on the movement of the binding constraint (5.4') for different iterations. Therefore the optimal solution and hence the optimal value Z_1 depend on the convergence of the constraint (5.4') where the optimal solution will be defined by the exact gradients λ_1, λ_2 of the response curves of extraction and injection cases correspondingly.

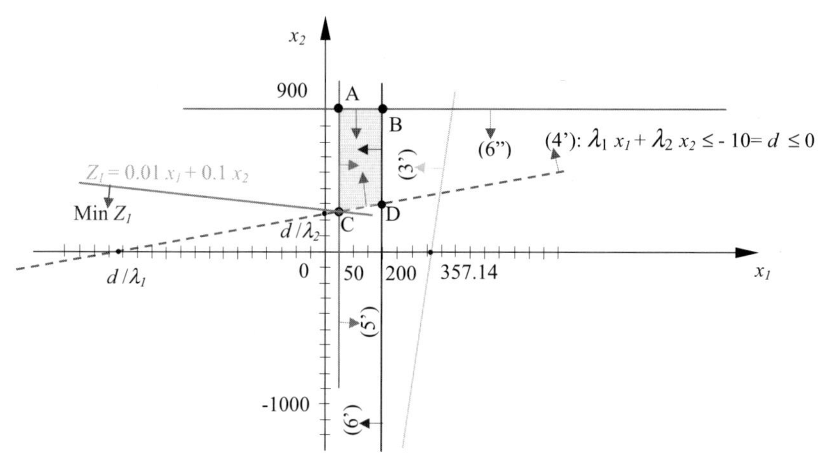

Figure 5.10: Graphical representation of the first objective function and its optimal solution at point *C*.

In Figure 5.10 it can be seen that the optimal solution, $x_1 = Q_{extract} = 50$, is constant after the first iteration since the initial values of x_1 and x_2 are set equal to zero. Thus if it is known that the solution x_1 will be a constant, $x_1 = 50$, (in fact it is possible to know through the constraint (5') by setting $x_1^0 = w_d = 50$) the initial value $x_1^0 = 50$ would be set and then the optimal solution x_1 would be fixed at the beginning of the first iteration. Therefore the value of λ_1^k, being the non-linear function $f^k(x_1)$, can be fixed at the first iteration, i.e. $\lambda_1^k(50) = \frac{\Delta L^k}{\Delta Q}(x_1 = 50) = f(50) = const$. The changes take place only on λ_2^k and then x_2^k. From (5.4') comes the inequality:

$$\lambda_1^k x_1^k + \lambda_2^k x_2^k \le -10 \tag{5.41}$$

and hence

$$\lambda_2^k x_2^k \le -10 - 50\lambda_1^1(x_1 = 50) = \text{constant} < 0,\ k = 2,3,\ldots, k_c. \tag{5.42}$$

Thus the left-hand side value of constraint (5.41) only depends on the term $\lambda_2 x_2$ at (or after) the first iteration (for our convenience $x_1^0 = 50$ is chosen). However, because the problem is minimizing Z_1 (to the southwest in the decision space) and constraint (5.41) is a binding constraint then the value of the left hand side of (5.41) must be equal to –10 (a condition for a binding constraint). This means that Z_1 will get the optimal value at the *C* extreme point. Consequently, the value of $\lambda_2 x_2$ will equal d – constant = –10 – constant = constant < 0 from the first iteration onward. That means when λ_2 is big, x_2 will be small and vice versa. Since d, λ_2 are negative and x_2 is positive, therefore λ_2 is simply mentioned as its absolute value.

On the other hand one recalls that λ_2 is the gradient of the response curve in the Figure 5.11 which is depicted for only one injection well in case of the quasi-three-dimensional model of the sensitivity analysis. The gradient λ_2 of the curve will increase when $x_2 = Q_{\text{inject}}$ increases, therefore the solution of the linearization approach will converge after several trial values of x_2 and λ_2. The Figures 5.12, 5.13, 5.14 show us how the convergence behaves when the upper bound and lower bound of x_2 are gradually closer to each other and the oscillation damping takes place.

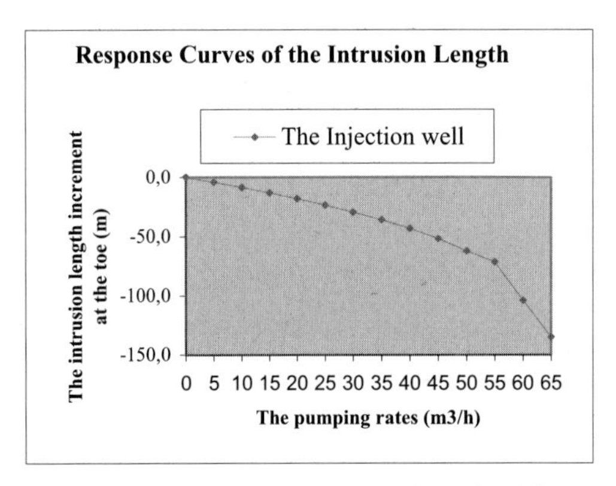

Figure 5.11: The response curve of the intrusion length with respect to injection rates.

The perturb pumping rate, ΔQ, is constant

Figure 5.12 shows the first iteration where $\lambda_2^1 = \lambda_2^{min}$ then by $\lambda_2^k x_2^k$ = const in the constraint (5.42) x_2^1 will get the maximum value, x_2^{max}. The LHS^1 and $Lsim^1$ stand for, correspondingly, the left hand side value of constraint (5.42) and the SW intrusion length simulated by SHARP relating to only the injection well in the first iteration.

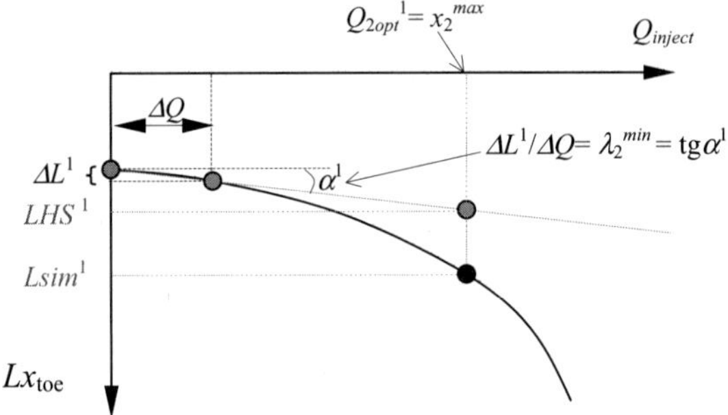

Figure 5.12: Schematic diagram of the iterative solution for finding the optimal solution.

In the second iteration, shown in Figure 5.13, the optimal solution $x_2^2 = Q_{opt}^2$ is obtained with the minimum value i.e., x_2^{min}, since $\lambda_2^2 = \lambda_2^{max}$.

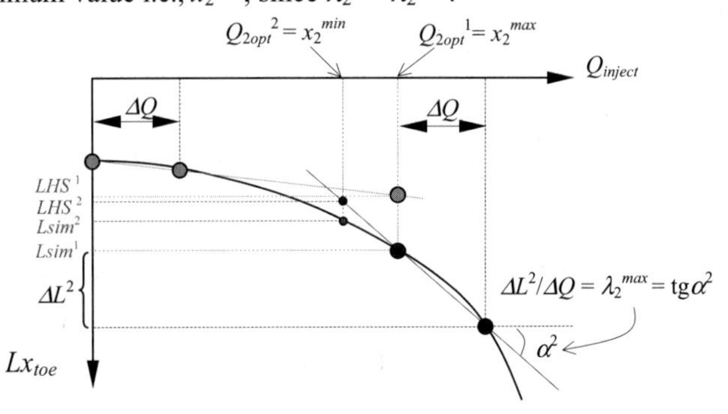

Figure 5.13: Schematic diagram of the iterative solution for finding the optimal solution continued

In the third iteration, shown in Figure 5.14, the optimal solution, $x_2^3 = Q_{2opt}^3$, which must be in-between the upper bound x_2^{max} and the lower bound x_2^{min}, will together with Q_{2opt}^2 set the new solution interval for the problem in the next iteration. It is because $\lambda_2^1 = \lambda_2^{min} \le \lambda_2^3 \le \lambda_2^2 = \lambda_2^{max}$ therefore $x_2^2 = x_2^{min} \le x_2^3 \le x_2^1 = x_2^{max}$.

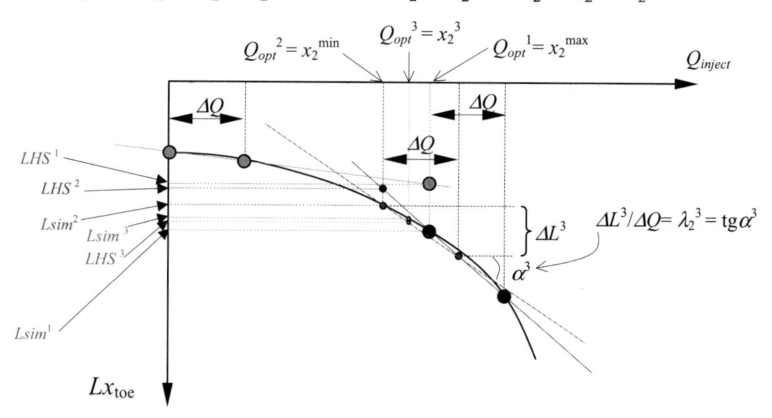

Figure 5.14: Schematic diagram of the iterative solution for finding the optimal solution continued.

If we continue with the fourth iteration, the fourth optimal solution $x_2^4 = Q_{2opt}^4$ will be obtained within the interval (x_2^2, x_2^3) since if λ_2^4 satisfies the interval $\lambda_2^3 \le \lambda_2^4 \le \lambda_2^2 = \lambda_2^{max}$ then $x_2^2 = x_2^{min} \le x_2^4 \le x_2^3$. We say "if " here because numerically we are not sure that the response curve is so smooth that we would assume the point in between two adjacent points has its gradient value inside the interval bounded by the gradient values of those two adjacent points e.g., $\lambda_2^{k+1} \le \lambda_2^{k+2} \le \lambda_2^k$. For example, in Table 5.1, for the 4th iteration the gradient λ_2^4 satisfies the bounded interval $\lambda_2^3 \le \lambda_2^4 \le \lambda_2^2$ (here we refer to their absolute values) hence $x_2^2 \le x_2^4 \le x_2^3$, but for the 5th iteration the gradient λ_2^5 does not satisfy the interval $\lambda_2^{k+1} \le \lambda_2^{k+2} \le \lambda_2^k$. This is because not only $\lambda_2^5 \ge \lambda_2^4$ but also $\lambda_2^5 \ge \lambda_2^3$. However, λ_2^5 satisfies the other bounded interval as in $\lambda_2^4 \le \lambda_2^5 \le \lambda_2^2$; therefore the solution x_2^5 satisfies the interval $x_2^2 \le x_2^5 \le x_2^4$ and it is ready for the 6th iteration.

Table 5.1: The convergence of the problem when $\Delta Q = 10$ m^3/h.

Iteration k	λ_1	λ_2	$x_1 = Q_{ext}$ (m^3/h)	$x_2 = Q_{inj}$ (m^3/h)	$\lVert Q_j^k - Q_j^{k-1} \rVert$	LHS ($=\lambda_2 Q_{inj}$) (m)	$Lsim$ (m)	Z_1 (MU/h)	$\lambda_1 x_1 + \lambda_2 x_2$
1	0.058848	-0.840376232	50	15.400738	10^5	-12.94241	-13.8125	2.0400	-10
2	0.058848	-1.045648742	50	12.377402	3.023	-12.94241	-10.9453	1.7377	-10
3	0.058848	-0.997793891	50	12.971030	0.593	-12.94241	-11.5234	1.7971	-10
4	0.058848	-1.003812559	50	12.893258	7.10^{-2}	-12.94241	-11.4375	1.7893	-10
5	0.058848	-1.004548429	50	12.883813	$9.\ 10^{-3}$	-12.94241	-11.4219	1.7883	-10
6	0.058848	-1.004404498	50	12.885660	$2.\ 10^{-3}$	-12.94241	-11.4297	1.7885	-10
7	0.058848	-1.004432641	50	12.885299	$3.\ 10^{-4}$	-12.94241	-11.4297	1.7885	-10
8	0.058848	-1.004427140	50	12,885369	$7.\ 10^{-5}$	-12.94241	-11.4297	1.7885	-10
9	0.058848	-1.004428207	50	12,885355	$1.\ 10^{-5}$	-12,94241	-11.4297	1.7885	-10
10	0.058848	-1.004427994	50	12,885358	$3.\ 10^{-6}$	-12,94241	-11.4297	1.7885	-10

Generally, the sequence goes on if after each iteration k the difference $\|x_2^k - x_2^{k-1}\| = \|Q_{2opt}^k - Q_{2opt}^{k-1}\|$ is bigger than ε. Otherwise it stops for the final solution $x_2^k = Q_{2opt}^k$. In Table 5.1 the value of $\|Q_{2opt}^k - Q_{2opt}^{k-1}\|$ decreases to 10^{-6} when the iteration varies from 1 to 10.

Numerically, our first objective optimization problem will converge to the optimal solution ($x_1 = Q_{ext} = 50$, $x_2 = Q_{inj} = 12.885$) after several iterations. In the table in which the computed values of λ_1, λ_2 are listed and so are the iterative optimal solution $x_1 = Q_{ext}$, $x_2 = Q_{inj}$, the *LHS* value of the constraint (5.42), the SW intrusion decrement (*Lsim*) and the optimal values Z_1.

In fact, one can stop the trial solution at the end of iteration 6 since the solution $x_2^6 = 12.885$ m^3/h is already exact to 3 decimal digits when compared with the last ones. On the other hand, one can also pose the question which is why bother with the convergence of the solution while in the practical field there is little difference between even 15 m^3/h and 12 m^3/h for installation and operation of a well. Here is just a simple example to show how the convergence takes place for such a problem or a similar one. For a more complicated problem with many more candidate wells, the algorithm can help with determining the wells at right locations and allocating the appropriate pumping rates, based on the converged optimal solution which can be very different indeed from the initial solution.

Decreasing the perturbed pumping rate

By this observation, we see that the convergence of the problem will be speeded up by gradually decreasing ΔQ for the iterations as shown in Table 5.2. In this problem, if we decrease ΔQ by 0.2 at every iteration step then the convergence of the solution takes place at the iteration 3 and the difference between the solutions between the two last iterations, $\|Q_j^k - Q_j^{k-1}\|$ is equal to zero. This means the optimal solution obtained from Table 5.2 is more accurate than the solution obtained from Table 5.1.

Table 5.2: The convergence of the problem with decreasing the perturbed pumping rate.

Iteration k	ΔQ	λ_1	λ_2	$x_1 = Q_{ext}$ (m^3/h)	$x_2 = Q_{inj}$ (m^3/h)	$\|Q_j^k - Q_j^{k-1}\| =$	*LHS* (=$\lambda_2 Q_{inj}$) (m)	*Lsim* (m)	Z_1 (MU/h)	$\lambda_1 x_1 + \lambda_2 x_2$
1	10	0.058800	−0.840376232	50	15.400738	10^4	−12.93662	−13.8125	2.040074	−10
2	9.8	0.058855	−1.043923171	50	12.398192	3.0025	−12.94276	−10.9609	1.739819	−10
3	9.6	0.058855	−1.043923171	50	12.398192	0.0	−12.94276	−10.9609	1.739819	−10

The multi-objective problem

For the multi-objective problem with two objectives as Z_1 and Z_2, if we use the weighting method with $w \geq 0$ then the objective $Z(w) = Z_1 + wZ_2$ has its gradient vector being just the linear combination of the gradient vectors of Z_1 and Z_2. That means the gradient vector of Z(w) lies inside the "cone" defined by the two gradient vectors of Z_1 and Z_2 as shown in Figure 5.15.

Recall that the weight w is the ratio of w_2/w_1 and that $Z(w)$ can be rewritten as $w_1 Z_1 + w_2 Z_2$. It can be seen that in Figure 5.15 the graph of $Z(w)$ is drawn clockwise from $Z(w = 0) = Z_1$ with $w_2 = 0$ to $Z(w = +\infty) = Z_2$ with $w_1 = 0$. In this case the $\mathbf{N}_d$ non-inferior set of Z_w is the vertical line segment AC, which is drawn in the decision space, with the coordinates of upper end point being (50, 900) and the lower end point being $(50,\ 1/\lambda_2(d-50\lambda_1))$. Please note that with a certain value of w, a so-called "switching value" (see Cohon, 1978), the weighted objective is parallel to the non-inferior set and then the weighted problem will have the alternative optima which are the points on the line segment AC. Therefore, in this problem the minimization of the objective $Z(w) = Z_1 + wZ_2$ will obtain the optimal solution either at the point C $(50,\ 1/\lambda_2(d-50\lambda_1))$ or A (50, 900) when $w = w_2/w_1$ varies in the range of $[0, +\infty)$. That means that when $w_1 = 0$ (or $w = \infty$) the tradeoff solution of the weighted problem will be the fixed point, A (50, 900) which is the same solution as in the single objective problem of Z_2 regardless of λ_j. When $w_2 = 0$ (or $w = 0$) the tradeoff solution of the weighted problem will then be the point C $(50,\ 1/\lambda_2(d-50\lambda_1))$, which depends on the

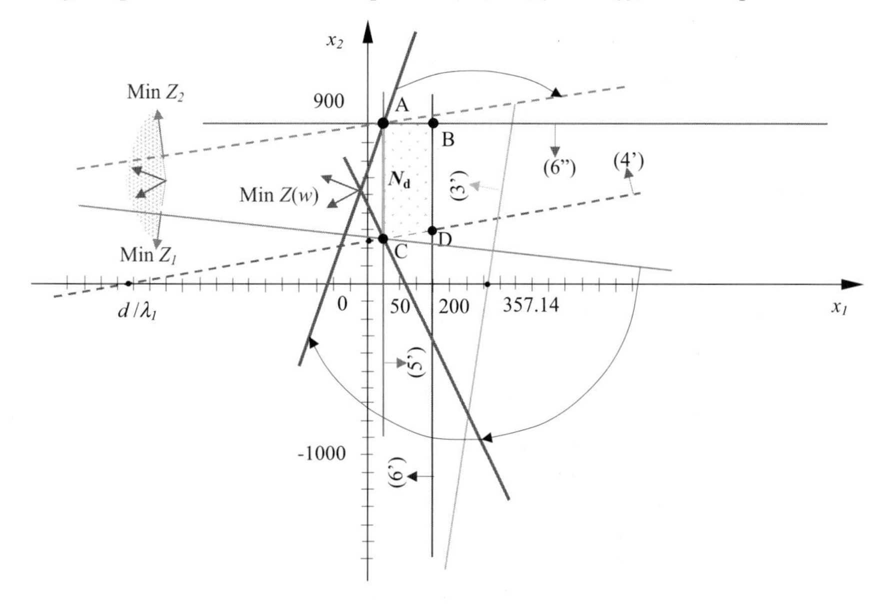

Figure 5.15: The feasible region and the multi-objective functions by the weighting method.

different iterations. The results of the latter case in terms of convergence of the optimal solution are shown in Table 5.3 and the non-inferior set in the objective space of the multi-objective problem for all iterations, when $w = 0$, is drawn in Figure 5.16 for the first iteration and in Figure 5.17 for the second iteration and the last iteration.

Table 5.3: The trade-off values (Z_1, Z_2) when $Z(w = 0) = Z_1$, and its solution x.

Iteration	ΔQ (m^3/h)	λ_1	λ_2	$x_1 = Q_{ext}$ (m^3/h)	$x_2 = Q_{inj}$ (m^3/h)	$Z_2 = \lambda^T x = \lambda_1 Q_1 + \lambda_2 Q_2$ (m)	$Z_1 = c^T x = 0.01Q_1 + 0.1Q_2$ (MU/h)	Lsim (m)
1	10	0.058800	-0.84037623	50	15.400738	-9,996619920	2,040074	-13,8125
2	9.8	0.058855	-1.04392317	50	12.398192	-10,000000007	1,739819	-10,9609
3	9.6	0.058855	-1.04392317	50	12.398192	-10,000000007	1,739819	-10,9609

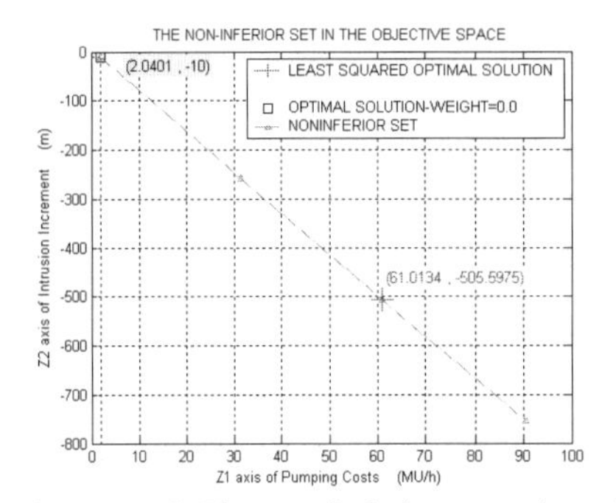

Figure 5.16: The non-inferior set at iteration 1.

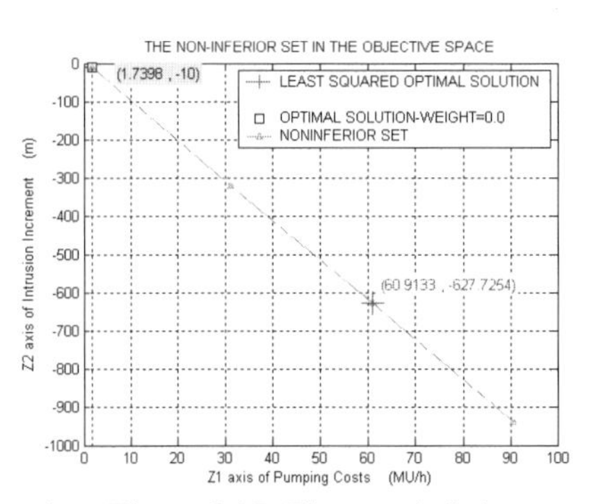

Figure 5.17: The non-inferior set at iterations 2 and 3.

5.4 Stochastic Optimization problem for SW intrusion management

5.4.1 Quadratic Cone Robust Single-Objective Optimization

Recall that in the previous sections the problem with the deterministic case in which the data are definitely known or assumed known. Here an LP program is also considered:

$$c^T x \to \text{Min} \;\big|\; Ax - b \geq 0 \tag{5.43}$$

However when such a problem arises in applications, its data *c, A, b* are not always known exactly; what is typically known is a domain **U** in the space of data – an "uncertainty set"- which definitely contains the "actual" unknown data. On the other hand, there are situations in reality where the design variable *x* must satisfy the "actual" constraints, whether or not they are known. In this situation, the only way to meet the requirements is to restrict oneself with robust feasible candidate solution *x* – those satisfying all possible realizations of the uncertain constraints, i.e. such that

$$Ax - b \geq 0 \;\; \forall [A;b] \;\; \exists c : (c, A, b) \in \mathrm{U} \tag{5.44}$$

To aggregate various realizations of the objective in something single, the same worst-case scenario and "guaranteed" objective *f(x)*- the maximum is used, over all possible realizations of the objective *c*, value of $c^T x$:

$$f(x) = \sup\left\{ c^T x \;\middle|\; c : \exists [A;b] : (c, A, b) \in \mathrm{U} \right\} \tag{5.45}$$

Then the robust feasible solution with the smallest possible value of the above guaranteed objective, so called the robust counterpart, are sought out, which is the optimization problem:

$$t \to \text{Min} \;\big|\; c^T x \leq t, \;\; Ax - b \geq 0 \;\; \forall (c, A, b) \in \mathrm{U} \tag{5.46}$$

However (5.46) is not an LP program because its structure depends on the geometry of the uncertainty set **U** and can be very complicated. Therefore, in many cases the

uncertainty set is specified as an ellipsoid - the image of the unit Euclidian ball under an affine mapping. Thus robust linear programming with ellipsoidal uncertainty sets can be viewed as a generic source of conic quadratic problems (see Ben-Tal and Nemirovski, 1998, Ndambuki, 2001).
Here in the problem it will be assumed that the uncertainty only affects the parameters of c and A, so the uncertain LP program is

$$\left\{ \begin{matrix} c^T x \to \text{Min} \\ a_i^T x - b_i \geq 0, \ i = 1, \ldots, m \end{matrix} \right\}_{(c,A) \in \mathcal{U}} \tag{5.47}$$

The uncertainty set is assumed to be:

$$\mathcal{U} = \left\{ (a_1; \ldots; a_m; c) \middle| \ c = \bar{c} + P_0 u_0, \ a_i = \bar{a}_i + P_i u_i, \ i = 1, \ldots, m, \text{ where } u_i^T u_i \leq 1 \right\}$$

where $\bar{c}, \bar{a}_i$ are the nominal data and $P_i u_i$, $i = 0, 1, \ldots, m$, represent the data perturbations and restrictions $u_i^T u_i \leq 1$ enforces these perturbations to vary in ellipsoids.

Thus the functional constraints and objective function will have the forms respectively:

$$\text{Min } t \ \left| \ \bar{c}^T x + u_0^T P_0^T x \leq t \text{ and } \bar{a}_i^T x + u_i^T P_i^T x - b_i \geq 0 \right. \tag{5.48}$$

Note that x is robust feasible if, and only if, for every $i = 1, \ldots, m$ there is

$$\underset{u_i : u_i^T u_i \leq 1}{\text{Min}} \left[\bar{a}_i^T x + u_i^T P_i^T x - b_i \right] \geq 0$$

or

$$\bar{a}_i^T x - b_i + \underset{u_i : u_i^T u_i \leq 1}{\text{Min}} u_i^T P_i^T x \geq 0$$

which is equivalent to (see Ben-Tal and Nemirovski, 1998; Ghaoui and Lebret, 1997)

$$\bar{a}_i^T x - b_i - \left\| P_i^T x \right\|_2 \geq 0$$

It is similar for the objective function:

$$\underset{u_0 : u_0^T u_0 \leq 1}{\text{Max}} \left[\bar{c}^T x + u_0^T P_0^T x - t \right] \leq 0$$

or

$$\bar{c}^T x - t + \underset{u_0 : u_0^T u_0 \leq 1}{\text{Max}} u_0^T P_0^T x \leq 0$$

which is equivalent to

$$\bar{c}^T x + \left\| P_0^T x \right\|_2 \leq t$$

From now on, the robust counterpart (5.46) becomes the quadratic cone program or the so-called second order cone (SOCO):

$$t \to \text{Min} \ \left| \ \bar{c}^T x + \left\| P_0^T x \right\| \leq t; \ \bar{a}_i^T x - \left\| P_i^T x \right\|_2 \geq b_i \ , \ i = 1, \ldots, m \right. \tag{5.49}$$

5.4.2 Robust single objective saltwater intrusion management

For the first application of the quadratic cone programming, every single objective function with the same number of constraints will be solved by assuming that the operational system is a linear scalar-optimization problem. The single objective function is either the minimization of the SW intrusion length or the minimization of costs incurred during the strategic operation within the allowable range of either the lowering of the hydraulic head or the increment/decrement of the SW intrusion length. For the single objective, Z_2, the deterministic optimization problem was:

$$\underset{x}{\text{Min}}\ [Z_2 = \lambda^T x], \tag{5.2}$$

subject to

$$\sum_{j=1}^{N_{ew}+N_{iw}} A_{i,j} x_j \geq b_i\ , \qquad i = 1,\dots,N_c \tag{5.3}$$

$$\sum_{i=1}^{N_{ew}+N_{iw}} \lambda_i x_i \leq d\ , \tag{5.4}$$

$$e^T x \geq W_D\ , \qquad e^T = (1_1,\dots,1_{New}) \tag{5.5}$$

$$x_i \leq Wc_i\ , \qquad i = 1,\dots,N_{iw}+N_{ew} \tag{5.6}$$

$$x_i \geq 0\ , \qquad i = 1,\dots,N_{iw}+N_{ew} \tag{5.7}$$

where those above terms are previously defined.

This deterministic optimization problem is transformed into the quadratic cone (QC) problem as follows:

$$\text{Min}\ (\bar{\lambda}^T x + \|P_0^T x\|_2), \tag{5.50}$$

subject to

$$\bar{a}_i^T x - \|P_i^T x\|_2 \geq b_i\ , \qquad i = 1,\dots,N_c \tag{5.51}$$

$$\bar{\lambda}^T x + \|P_0^T x\|_2 \leq d\ , \tag{5.52}$$

$$e^T x \geq W_D\ , \tag{5.5}$$

$$x_i \leq Wc_i\ , \qquad i = 1,\dots,N_{iw}+N_{ew} \tag{5.6}$$

$$x_i \geq 0\ , \qquad i = 1,\dots,N_{iw}+N_{ew} \tag{5.7}$$

in which

$$(a_i;\lambda) \in \mathbf{U} = \{\bar{a}_i + P_i u\ ; \bar{\lambda} + P_0 u\ \big|\ \|u\| \leq 1\} \tag{5.53}$$

where P_0, P_i are $n \times \xi$ matrices of perturbations, which describe the shape and size of the uncertainty in the response λ, a_i, where $u \in R^{\xi} : \|u\| \leq 1$ enforces the perturbation to vary in the ellipsoids, and $\bar{a}_i, \bar{\lambda} \in R^n$ are the nominal responses and forms the center of the ellipsoids defined by Equation (5.53). The norm terms in (5.51), (5.52) translate into $(x^T P_i P_i x)^{1/2}$ and $(x^T P_0 P_0 x)^{1/2}$. Thus $D_i = P_i P_i$ and $D_0 = P_0 P_0$ are the covariance matrices.

Since our objective function on the SW intrusion length increment is non-linear, thus it has to be transformed into the linear one by using a variable t as follows:

$$\text{Min} \quad t, \tag{5.54}$$

subject to

$$\bar{\lambda}^T x + \left\| P_0^T x \right\|_2 \le t, \tag{5.55}$$

$$\bar{a}_i^T x - \left\| P_i^T x \right\|_2 \ge b_i, \quad i = 1,\dots,N_c \tag{5.51}$$

$$\bar{\lambda}^T x + \left\| P_0^T x \right\|_2 \le d, \tag{5.52}$$

$$e^T x \ge W_D, \tag{5.5}$$

$$x_i \le Wc_i, \qquad i = 1,\dots,N_{iw}+N_{ew} \tag{5.6}$$

$$x_i \ge 0, \qquad i = 1,\dots,N_{iw}+N_{ew} \tag{5.7}$$

The robust single objective problem has the form that can be solved by the conic quadratic programming with the new variable (t, x) and the former objective (5.50) becomes the constraint (5.55). Recall that in the epigraph of $g(x)$, the inequality constraints (5.55), (5.51) and (5.52) are similar to the CQ-representable function of the Euclidean norm., i.e., the epigraph of g is given by the conic quadratic inequality of $\|x\|_2 \le t$. The linear inequality constraints (5.5), (5.6), (5.7), which are in the forms of CQ-representable functions of linear functions, can be translated into the conic quadratic forms as $\|0\|_2 \le p^T z - q$ where $z = (t, x)$. Hence, the whole problem can be formulated under the dual standard form of mixed quadratic cone (QC) and the cone of nonnegative orthant. The cone $\mathbf{K}$ now is comprised of $\mathbf{R}^m_+ \times \mathbf{Q}_{\text{cone}} \times \mathbf{Q}_{\text{cone}} \times \mathbf{Q}_{\text{cone}}$ as:

$$\text{Max} \quad \begin{bmatrix} -1 & 0 \end{bmatrix} \cdot \begin{bmatrix} t \\ x \end{bmatrix},$$

s.t.

$$(5.5) \Leftrightarrow \quad -W_D + e^T x \ge 0,$$

$$(5.6) \Leftrightarrow \quad Wc_i - x_i \ge 0,$$

$$(5.7) \Leftrightarrow \quad x_i \ge 0,$$

$$(5.55) \Leftrightarrow \quad t - \bar{\lambda}^T x \ge \left\| P_0^T x \right\|_2 \Leftrightarrow \left(t - \bar{\lambda}^T x, \left\| P_0^T x \right\|_2 \right) \in Q_{cone},$$

$$(5.51) \Leftrightarrow \quad -b_i + \bar{a}_i^T x \ge \left\| P_i^T x \right\|_2 \Leftrightarrow \left(-b_i + \bar{a}_i^T x, \left\| P_i^T x \right\|_2 \right) \in Q_{cone},$$

(5.52) $\Leftrightarrow$ $d - \bar{\lambda}^T x \geq \|P_0^T x\|_2 \Leftrightarrow \left(d - \bar{\lambda}^T x, \|P_0^T x\|_2\right) \in Q_{cone}$,

For solving the problem with SeDuMi, refer to Chapter 4.

For the single objective problem with $Z_1 = c^T x$, being the linear objective, it is not necessary to introduce the variable t and for the constraints the same procedure will be carried out as in the single objective problem of Z_2. Thus:

Max $-c^T x$,

s.t.

(5.5) $\Leftrightarrow$ $-W_D + e^T x \geq 0$,

(5.6) $\Leftrightarrow$ $Wc_i - x_i \geq 0$,

(5.7) $\Leftrightarrow$ $x_i \geq 0$,

(5.51) $\Leftrightarrow$ $-b_i + \bar{a}_i^T x \geq \|P_i^T x\|_2 \Leftrightarrow \left(-b_i + \bar{a}_i^T x, \|P_i^T x\|_2\right) \in Q_{cone}$,

(5.52) $\Leftrightarrow$ $d - \bar{\lambda}^T x \geq \|P_0^T x\|_2 \Leftrightarrow \left(d - \bar{\lambda}^T x, \|P_0^T x\|_2\right) \in Q_{cone}$,

In this problem the cone **K** is comprised of $\mathbf{R}^m_+ \times \mathbf{Q}_{\text{cone}} \times \mathbf{Q}_{\text{cone}}$. In order to solve the problem with SeDuMi, refer to Chapter 4.

5.4.3 The uncertainty set, U, of the robust linear programming for SW intrusion management

It is known that in the robust linear programming the data (c, A, b) belong to the "uncertainty set" **U** which must be defined and computed before the optimization program will be executed. In this problem the data coefficients c and A which are mostly influenced by the uncertainty of the hydraulic conductivity are proposed. Although those coefficients can be influenced by the other factor e.g., the boundary flux, as well (see the sensitivity analysis) here in this problem the uncertainty of hydraulic conductivity is considered more important and it needs to be studied.

Generating the realizations of the hydraulic conductivity

Since the exact value of the hydraulic conductivity, *k,* are not knownin every cell of the aquifer, the random values of *k* will be generated for all cells of the aquifer. This can be modelled with the **S**equential **G**aussian **Sim**ulation (SGSIM) in the GSLIB software (Deutsch and Journel, 1998). The parameter values of the model are defined in the parameter input file, *sgsim.par* in which the kriging variance of the conditional distribution, *cc*, and the isotropic nugget constant *co* are given. The other important parameters of the model are the discretization information of the three-dimension aquifer i.e., *nx*, *xsiz*, *ny*, *ysiz*, *nz*, *zsiz*, which are the number and size of cells in the *x*, *y*, and *z* directions, respectively. The number of simulations to generate, *nsim*, is also assigned to the parameter input file (see Deutsch and Journel, 1998.)

For example, with the parameters *nsim* = 20, *cc* = 0.9, *co* = 0.1, *nx* = 66, *xsiz* = 1250, *ny* = 15, *ysiz* = 1250, *nz* = 1, *zsiz* = 151, the two-dimensional distribution of the hydraulic conductivity for the first realization in Figure 5.18 and the histogram for all random data in Figure 5.19 are as given below:

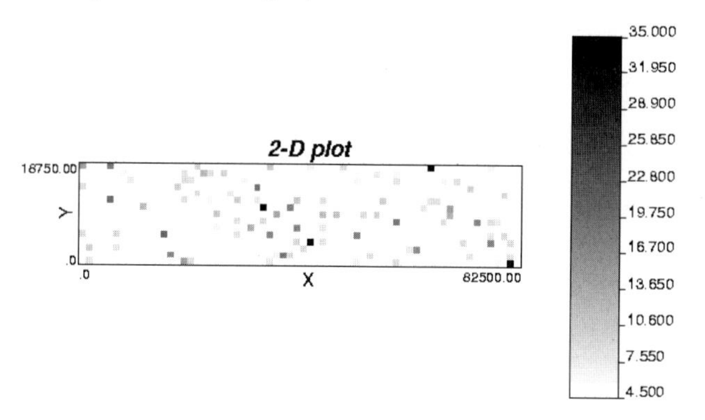

Figure 5.18: The 2-D plan view for the first realization of hydraulic conductivity.

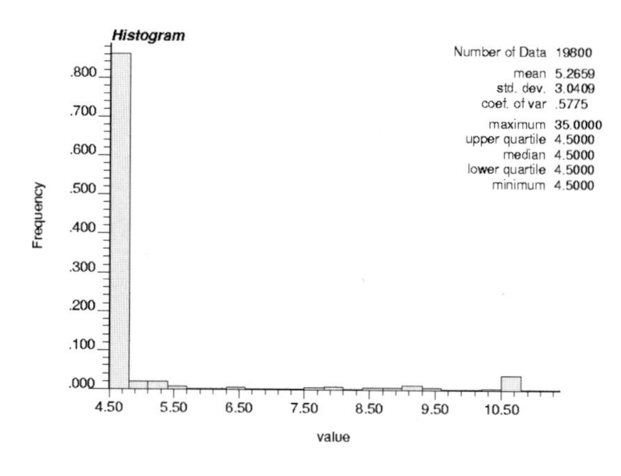

Figure 5.19: Histogram of 20 realizations of the generated hydraulic conductivity.

Computing the nominal vectors and perturbation matrices

Given ξ number of realizations of hydraulic conductivity values, the ξ response head matrices $A_{ij} = [\Delta H_i / \Delta Q_j]$, $i = 1,\ldots, N_c$ and $j = 1,\ldots, N_w$ and the ξ response intrusion length vectors $\lambda_j = [\Delta L / \Delta Q_j]$, $j = 1,\ldots, N_w$, are correspondingly computed by running SHARP computer code ξ times. In the robust linear programming the data $(c^T, A, b) \in$ **U**, so-called "uncertainty set", need to be defined. Here, in this problem the vector c^T, whose components are the coefficients in the objective Z_2 and the constraint (5.52), is defined by the nominal vector $\overline{\lambda}^T$ and the perturbation matrix P_0^T due to the response of salt/freshwater interface. Also the matrix A whose components are mainly the coefficients of the constraint (5.51) is defined by the nominal vectors $\overline{a}_i^T$ and the perturbation matrices P_i^T , $i = 1,\ldots, N_c$, due to the head response.

a. The nominal vector and perturbation matrix due to the response vector of the average SW intrusion length

The size of the response intrusion length vector is equal to $[N_w, 1]$ where N_w is the number of design variables, i.e. number of candidate wells. In total there are ξ observations for each variable i.e. the ξ random values of $\lambda_j = (\Delta L / \Delta Q_j)$, $j = 1, \ldots, N_w$, which are obtained by ξ realizations of hydraulic conductivity values.

The matrix arrangement of ξ realizations of the response intrusion length vector is shown in Table 5.4.

Table 5.4: The matrix of ξ realizations of the response intrusion length vector.

The Response Intrusion length (λ_{jk})		Realization number (or observation) ξ				
		1st	2nd	3rd	...	ξ^{th}
Design variable, j =	1	λ_{11}	λ_{12}	λ_{13}	...	$\lambda_{1\xi}$
	2	λ_{21}	λ_{22}	λ_{23}	...	$\lambda_{2\xi}$
	3	λ_{31}	λ_{32}	λ_{33}	...	$\lambda_{3\xi}$
	...	....	....	....	...	....
	n	λ_{n1}	λ_{n2}	λ_{n3}	...	$\lambda_{n\xi}$

Here a_1 is a vector of uncertain input parameters lying in a given ellipsoid, Ξ_1 based on the matrix $\{\lambda_{jk}\}_1$, (the uncertainty set) defined under a general form for every uncertain input parameter a_i as (see Ben-Tal and Nemirovski, 1998, Ndambuki, 2001)

$$\Xi_i = \left\{ a_i = \bar{a}_i + P_i u \;\middle|\; \|u\| \le 1 \right\}$$

Here P_i is an $N_w \times \xi$ perturbation matrix which describes the shape and size of the uncertainty in the response a_i, where the index i indicates the ordinal number of control points, $u \in R^{\xi}$: $\|u\| \le 1$ enforces the perturbations to vary in ellipsoids, and $\bar{a}_i \in R^n$, $n = N_w$, are the nominal responses and forms the center of the ellipsoids defined by the above equation. Furthermore, Ξ_i can be considered as full dimension ellipsoids with the additional requirement of invertibility of P_i with $\xi = n$ and with the property that $P_i = P_i^T$ and P_i is a positive definite matrix (see Ndambuki, 2001).

The nominal vector $\bar{\lambda}_j$ of the matrix λ_{jk} is defined as the arithmetic mean of every row j of the λ_{jk} matrix, e.g. in the Matlab script it can be obtained by the command:

```
>>Lxtoemat = mean(λ,2)
```

The size of $\bar{\lambda}_j$ vector is $[n, 1]$.

The perturbation matrix P_0^T of the matrix λ_{jk} is defined as the covariance matrix C_{jq} ($j, q = 1, \ldots, n = N_w$) of the matrix $(\lambda_{jk})^T = \lambda_{kj}$ where k is the row of observations and j is the column of variables. Each component of the covariance matrix is defined as the covariance of each pair of row-vector variables, $(\lambda_{jk}, \lambda_{qk})$, where $j, q = 1, \ldots, n = N_w$. For example, if using the Matlab script the perturbation matrix can be obtained by the command:

```
>>cov(λ')
```

The size of $P_0^T = \text{cov}(\lambda_{jk}')$ is $[n, n]$.

b. The nominal vector $\bar{a}_i$ and perturbation matrix $\mathbf{P}_i$ of the response head matrix

For the response head matrix, if there are N_c control points (or N_c constraints) and n candidate wells (or n design variables), thus the size of the response head matrix $A_{ij} = [\Delta H_i/\Delta Q_j]$ is equal to $[N_c, n]$. If the robust linear problem is to be treated as a set of all single head constraints, there will be N_c head constraint inequalities. Each head constraint inequality has its coefficients being the components of each row vector of A_{ij} (with the size of $[1, n]$). In the uncertainty problem, if there are ξ observations for each variable, j, e.g. the ξ random values of each row of A_{ij} which are obtained by ξ realizations of hydraulic conductivity values, then there will be the results in Table 5.5. This shows the matrix arrangement of ξ realization of the response head vector at each control point (for each constraint).

Table 5.5: The matrix of ξ realization of the response head vector at each control point

The Response Head at one control point (A_{jk})		Realization number (or observations) $k = 1,\dots, \xi$				
		1st	2nd	3rd	...	ξ [th]
Design variable, j =	1	A_{11}	A_{12}	A_{13}	...	$A_{1\xi}$
	2	A_{21}	A_{22}	A_{23}	...	$A_{2\xi}$
	3	A_{31}	A_{32}	A_{33}	...	$A_{3\xi}$
	...	....	....	....	...	....
	n	A_{n1}	A_{n2}	A_{n3}	...	$A_{n\xi}$

The nominal vector $\bar{a}$ of the response matrix A_{jk} at each control point is defined as the arithmetic mean of every row of the matrix, e.g. in the Matlab script it can be obtained by the command:

```
>>Hpmatrix = mean(A,2)
```

The size of $\bar{a}$ vector is $[n, 1]$.

The perturbation matrix P of the response matrix A_{jk} at each control point (or constraint) is defined as the covariance matrix C_{jq} $(j, q = 1,\dots,n = N_w)$ of the matrix $(A_{jk})^T = A_{kj}$. Each component of the covariance matrix is the covariance of each pair of row-vector variables, (A_{jk}, A_{qk}), where $j, q = 1,\dots,n = N_w$.

Using the Matlab script the perturbation matrix can be obtained by the command

```
>>cov(A')
```

The size of $P = \text{cov}(A')$ is $[n, n]$.

c. The level of robustness

In order to achieve a different level of uncertainty or robustness Ndambuki (2001) proposed in his work the scaling factor η that is multiplied to the perturbation matrices. Here also in this thesis the same η factor is used for both perturbation matrices as P_0 and P_i. Thus the perturbation matrices as P_0 and P_i are replaced by ηP_0 and ηP_i, respectively. The value of η varies from 0 to greater than 1, when $\eta = 0$ that means there is no uncertainty (the parameters are exactly known), $\eta < 1$ means that the uncertainty level is less than that given by the perturbation matrices, $\eta = 1$ means that the uncertainty is as given by the perturbation matrices, and $\eta > 1$ means that the uncertainty is higher than that given by the perturbation matrices (see Ndambuki, 2001). From now on the robust single objective problem for SW intrusion management model is rewritten as:

$$\text{Min} \quad t \tag{5.54}$$

Subject to

$$\bar{\lambda}^T x + \left\| \eta P_0^T x \right\|_2 \leq t, \tag{5.55'}$$

$$\bar{a}_i^T x - \left\| \eta P_i^T x \right\|_2 \geq b_i, \quad i = 1,\dots,N_c \tag{5.51'}$$

$$\bar{\lambda}^T x + \left\| \eta P_0^T x \right\|_2 \leq d, \tag{5.52'}$$

$$e^T x \geq W_D, \tag{5.5}$$

$$x_i \leq Wc_i, \quad i = 1,\dots,N_{iw}+N_{ew} \tag{5.6}$$

$$x_i \geq 0, \quad i = 1,\dots,N_{iw}+N_{ew} \tag{5.7}$$

5.4.4. Solution methodology

To solve the above robust single objective SOCO problem (formulations (5.54), (5.55'), (5.51'), (5.52'), (5.5), (5.6) and (5.7)), there are the following steps:

2. Generating ξ number of realizations of the uncertain parameter (in this case, the aquifer hydraulic conductivity field). Each realization, ξ, has different spatial values of the hydraulic conductivity, k.
27. Based on the two-fluid groundwater flow simulation software, SHARP, and the different realizations of hydraulic conductivity fields, the intrusion length response matrix λ_{jk} and the groundwater head response matrices A_{jk} are computed for each control point.
28. The nominal vectors, $\bar{\lambda}, \bar{a}_i$ and the perturbation matrices, ηP_0, ηP_i are computed.
29. Using the quadratic cone programming in SeDuMi, to solve the robust single objective problem.

5.4.5 Quadratic Cone Robust Multi-Objective Optimization

Each single-objective problem has been solved separately by application of the quadratic cone, or the so-called second order cone (SOCO), programming. With those single optimal values computed the payoff table is generated. In this table the solutions are incomparable or the values of the multi-objective problem are associated with the trade-off problem as long as these objectives, $Z_k(x)$, are competing from one to another. Otherwise there is little reason for setting up such objective functions in the optimization problem. This means that there is no unique solution to such a problem. As used in the multi-objective problem for the deterministic case, the two methods, i.e. the minimum distance from the ideal solution and prior assessments of weights (weighting method), are also proposed for this robust multi-objective optimization problem.

The minimum distance from the ideal solution

Ndambuki (2001) proposed application of the goal programming to solve such problems by specifying the aspiration level for each of the linear objectives and then seeking the solution that minimizes the sum of the deviations of those linear objective functions from their respective goals. If the aspiration levels of the objective function ($Z_k(x)$, by $\overline{Z}_k(x)$ for k=1,...K) are denoted then it results in the form of deviation, $d_k = \overline{Z}_k - Z_k$, for k=1,...K.

The above deviation can be expressed by the distance between two point vectors, $v_i, \overline{v}_i$, in the n dimension vector space set of $\mathbf{R}^n$. Its distance can be measured by the L_2 metric:

$$\|d\|_2 = \|v - \overline{v}\|_2 = \left(\sum_{k=1}^{n} |v_k - \overline{v}_k|^2 \right)^{\frac{1}{2}} \tag{5.56}$$

The L_2-metric will make the problem non-linear and can be solved as a quadratic optimization or the quadratic cone optimization (SOCO) problem.

However, as discussed in the deterministic problem, the optimal solution obtained by this method is not always in the non-inferior set where decision-makers can obtain the best-compromised solution as it should be. Therefore, it is proposed using the method of minimum distance from the ideal solution and consider $\overline{Z} = \{\overline{Z}_1,...,\overline{Z}_K\}^T$ as the ideal solution (utopia point) of $Z(x) = \{Z_1(x),...,Z_K(x)\}^T$ the distance between those two vectors can be also measured by the L_α-metric:

$$\|\overline{Z} - Z(x)\|_\alpha = \left(\sum_{k=1}^{K} |\overline{Z}_k - Z_k(x)|^\alpha \right)^{\frac{1}{\alpha}} \tag{5.57}$$

where α=1, 2,...,∞.

The multi-objective optimization problem has the form:

$$\min_{x \in \Omega} \|\overline{Z} - Z(x)\|_\alpha \tag{5.58}$$

where x is the decision variable, $\mathbf{\Omega}$ is the feasible set. By introduction of the scalar deviation variable, $t \in \mathbf{R}_+$, formula (5.58) will be translated to the form:

$$\text{Min} \quad t_\alpha \tag{5.59}$$

s.t.

$$\|\overline{Z} - Z(x)\|_\alpha \le t_\alpha, \tag{5.60}$$

$$x \in \mathbf{\Omega} \tag{5.61}$$

This robust objective constraint problem will be solved as the quadratic cone optimization (SOCO) problem with the new variable vector (t, x).

The prior assessments of weights

Another method that can be preferable to decision makers is the prior assessments of weights. This method that was discussed in the deterministic problem gives more choices to decision makers in the sense of stating their preferences among multiple objectives and ensuring the non-inferiority of the optimal solution when $w > 0$. Thus

$$\text{Min} \quad Z(w) = Z_1 + wZ_2 \tag{5.62}$$

s.t.

$$x \in \mathbf{\Omega} \tag{5.61}$$

Since some objectives can be of non-linear functions whose coefficients are defined in the "uncertainty set" **U,** then the objective of the weighted problem will be also non-linear and it is formulated as the robust multi-objective problem as follows:

$$\text{Min} \quad t_w \tag{5.63}$$

s.t.

$$Z_1 + wZ_2 \le t_w, \tag{5.64}$$

$$x \in \mathbf{\Omega} \tag{5.61}$$

and this can be solved by the quadratic cone programming as well.

5.4.6 Robust multi-objective saltwater intrusion management

In the previous section a multi-objective optimization problem with two-objective functions, minimization of the intrusion length and the total cost of strategic operation was presented. The robust multi-objective quadratic cone (SOCO) problem is as follows:

$$\underset{x}{\text{Min}} \quad (Z_1 = c^T x) \tag{5.1}$$

$$\underset{x}{\text{Min}} \ (Z_2 = \bar{\lambda}^T x + \left\|\eta P_0^T x\right\|_2) \tag{5.50'}$$

s.t.

$$\bar{a}_i^T x - \left\|\eta P_i^T x\right\|_2 \ge b_i, \quad i = 1,\dots,N_c \tag{5.51'}$$

$$\bar{\lambda}^T x + \left\|\eta P_0^T x\right\|_2 \le d, \tag{5.52'}$$

$$e^T x \ge W_D, \tag{5.5}$$

$$x_i \le Wc_i, \quad i = 1,\dots,N_{iw}+N_{ew} \tag{5.6}$$

$$x_i \ge 0, i = 1,\dots,N_{iw}+N_{ew} \tag{5.7}$$

Where Z_1 and Z_2 are the linear and non-linear objective functions, respectively.

The method of minimum distance from the ideal solution

With this method the robust multi-objective problem for the SW intrusion management has the form

$$\underset{x\in\Omega}{\text{Min}} \ \left\| \overline{Z} - Z(x) \right\|_{\alpha}, \qquad (5.58)$$

s.t.

$$x \in \mathbf{\Omega} \ , \qquad (5.61)$$

where $\overline{Z} = [\overline{Z}_1, \overline{Z}_2]^T$ is the ideal point, and $\overline{Z}_1$, $\overline{Z}_2$ are the optimal values of Z_1 and Z_2, respectively; $Z(x) = [Z_1(x), Z_2(x)]^T = [c^T x \ , \ \lambda^T x + \|P. \ x\|]^T$; $\mathbf{\Omega}$ is the feasible set defined by constraints (5.51'), (5.52'), (5.5), (5.6) and (5.7).

$\left\| \overline{Z} - Z(x) \right\|_{\alpha}$ is the L_{α} metric with:

$$\left\| \overline{Z} - Z(x) \right\|_{\alpha} = \left[\sum_{k=1}^{n} \left| \overline{Z}_k - Z_k(x) \right|^{\alpha} \right]^{1/\alpha} ; \alpha \in \{1,2,3,...\} \cup \{\infty\} \qquad (5.65)$$

Since the single objective problem is minimizing Z_k then it is possible to state with certainly that $\overline{Z}_k \leq Z_k$. So $|\overline{Z} - Z(x)|^{\alpha}$ can be substituted by $(Z_k(x) - \overline{Z}_k)^{\alpha}$.
The problem (5.58) can be rewritten as:

$$\text{Min} \quad t_{\alpha} \qquad (5.59)$$

or

$$\text{Max} \quad (-t_{\alpha}), \qquad (5.59')$$

s.t.

$$t_{\alpha} \geq \left\| \overline{Z} - Z(x) \right\|_{\alpha}, \qquad (5.60)$$

$$x \in \mathbf{\Omega} . \qquad (5.61)$$

For $\alpha = 1$, (5.60) becomes:

$$t_1 \geq \| \overline{Z}_k - Z_k \|_1 = \left[\sum_{k=1}^{2} \left(Z_k - \overline{Z}_k \right)^1 \right]^1 = (Z_1 - \overline{Z}_1) + (Z_2 - \overline{Z}_2)$$

$$= c^T x - \overline{Z}_1 + \lambda^T x + \|\eta P_0 \, x\| - \overline{Z}_2$$

or

$$t_1 - c^T x + \overline{Z}_1 - \lambda^T x + \overline{Z}_2 \geq \|\eta P_0 \, x\| \qquad (5.66)$$

Then the problem can be formulated as a conic quadratic program under the matrix form:

$$\text{Max} \qquad \begin{bmatrix} -1 & 0 \end{bmatrix} . \begin{bmatrix} t_1 \\ x \end{bmatrix},$$

s.t.

$$\begin{bmatrix} \overline{Z}_1 + \overline{Z}_2 \\ 0 \end{bmatrix} - \begin{bmatrix} -1 & (c^T + \lambda^T) \\ 0 & -\eta P_0 \end{bmatrix} . \begin{bmatrix} t_1 \\ x \end{bmatrix} \in Qcone$$

and $\quad x \in \mathbf{\Omega}$

For α = ∞, (5.60) becomes:

$$t_\infty \geq \| \overline{Z}_k - Z_k \|_\infty = \max_{k=1,\dots,n} \left\{ \left(Z_k - \overline{Z}_k \right) \right\} = \max \left\{ \left(Z_1 - \overline{Z}_1 \right), \left(Z_2 - \overline{Z}_2 \right) \right\}$$

Then the problem can be formulated as a conic quadratic program:

$$\text{Max} \quad (-t_\infty),$$

$$\text{s.t. } t_\infty \geq c^T x - \overline{Z}_1, \tag{5.67}$$

$$t_\infty \geq \lambda^T x + \|\eta P_0 x\| - \overline{Z}_2 \tag{5.68}$$

and $\quad x \in \mathbf{\Omega}$

where (5.67) is a linear constraint and (5.68) is rewritten under the conic quadratic form. The whole problem is written in the matrix form as:

$$\text{Max} \quad \begin{bmatrix} -1 & 0 \end{bmatrix} . \begin{bmatrix} t_\infty \\ x \end{bmatrix}$$

s.t.

$$\left[\overline{Z}_1 \right] - \begin{bmatrix} -1 & c^T \end{bmatrix} . \begin{bmatrix} t_\infty \\ x \end{bmatrix} \geq 0,$$

$$\begin{bmatrix} \overline{Z}_2 \\ 0 \end{bmatrix} - \begin{bmatrix} -1 & \lambda^T \\ 0 & -\eta P_0 \end{bmatrix} . \begin{bmatrix} t_\infty \\ x \end{bmatrix} \in Qcone$$

and $\quad x \in \mathbf{\Omega}$

For α = 2, the objective of $\underset{x \in \Omega}{\text{minimize}} \ \left\| \overline{Z} - Z(x) \right\|_2$ is formulated as:

$$\underset{x \in \Omega}{\text{Min}} \ \left\| \overline{Z} - Z(x) \right\|_{w,2} = \underset{x \in \Omega}{\text{Min}} \left(\sqrt{(Z_1(x) - \overline{Z}_1)^2 + (Z_2(x) - \overline{Z}_2)^2} \right)$$

$$= \underset{x \in \Omega}{\text{Min}} \left(\sqrt{(c^T x - \overline{Z}_1)^2 + (\overline{\lambda}^T x + \left\| \eta P_0^T x \right\|_2 - \overline{Z}_2)^2} \right) \tag{5.69}$$

which is equivalent to

$$\underset{x \in \Omega}{\text{Min}} \left((c^T x - \overline{Z}_1)^2 + (\overline{\lambda}^T x + \left\| \eta P_0^T x \right\|_2 - \overline{Z}_2)^2 \right) \tag{5.70}$$

or

$$\text{Max} \quad (-t_2) \tag{5.71}$$

s.t.

$$t_2 \geq (c^T x - \overline{Z}_1)^2 + (\overline{\lambda}^T x + \left\| \eta P_0^T x \right\|_2 - \overline{Z}_2)^2, \tag{5.72}$$

$$x \in \mathbf{\Omega}$$

since $\bar{\lambda}^T x + \|\eta P_0^T x\|_2 - \bar{Z}_2 > 0$ and if $\beta \geq \bar{\lambda}^T x + \|\eta P_0^T x\|_2 - \bar{Z}_2$ is set then (5.72) becomes

$$t_2 \geq (c^T x - \bar{Z}_1)^2 + \beta^2 \tag{5.73}$$

and

$$\beta \geq \bar{\lambda}^T x + \|\eta P_0^T x\|_2 - \bar{Z}_2 \tag{5.74}$$

Recalling the elementary QC-representable functions in Chapter 4, (5.73) is a form of the squared Euclidean norm and (5.74) is a form of the Euclidean norm, and (5.73) and (5.74) can be rewritten in the forms of quadratic cones, respectively, as:

$$\frac{t_2+1}{2} \geq \left\| \left(\frac{t_2-1}{2}, \beta, c^T x - \bar{Z}_1 \right)^T \right\|_2 \qquad (5.73')$$

and

$$\bar{Z}_2 + \beta - \bar{\lambda}^T x \geq \|\eta P_0^T x\|_2 \qquad (5.74')$$

The other constraints (5.51'), (5.52') and (5.5), (5.6), (5.7) can be translated into the conic quadratic forms as in the robust single objective problems.

Therefore the robust multi-objective problem in the minimum distance method from the ideal solution with L_2-metric will be formulated as below:

$$\text{Max} \quad t_2 \tag{5.71}$$

s.t.

$$\frac{t_2+1}{2} \geq \left\| \left(\frac{t_2-1}{2}, \beta, c^T x - \bar{Z}_1 \right)^T \right\|_2, \qquad (5.73')$$

$$\bar{Z}_2 + \beta - \bar{\lambda}^T x \geq \|\eta P_0^T x\|_2, \qquad (5.74')$$

$$-b_i + \bar{a}_i^T x \geq \|\eta P_i^T x\|_2, \quad i = 1,\dots,N_c \qquad (5.51')$$

$$d - \bar{\lambda}^T x \geq \|\eta P_0^T x\|_2, \qquad (5.52')$$

$$-W_D + e^T x \geq 0, \tag{5.5}$$

$$Wc_i - x_i \geq 0, \quad i = 1,\dots,N_{iw}+N_{ew} \tag{5.6}$$

$$x_i \geq 0, \quad i = 1,\dots,N_{iw}+N_{ew} \tag{5.7}$$

For solving this type of problem, refer to Chapter 4.

The method of prior assessments of weights

With a given weight, $w > 0$, the robust multi-objective problem for SW intrusion management has the form:

$$\text{Min} \quad t_w \tag{5.63}$$

s.t.

$$t_w \geq Z_1 + wZ_2 = c^T x + w\left(\bar{\lambda}^T x + \left\|\eta P_0^T x\right\|_2\right), \tag{5.75}$$

$$x \in \mathbf{\Omega} \tag{5.61}$$

where (5.75) is a form of CQ-representable function of Euclidean norm with the nonnegative weights. It will be formulated as follows:

$$t_w - (c^T + w\bar{\lambda}^T)x \geq w\left\|\eta P_0^T x\right\|_2 \tag{5.75'}$$

Together with the constraints (5.51'), (5.52'), (5.5), (5.6) and (5.7), the whole problem will be:

$$\text{Max} \quad -t_w$$

s.t.

$$t_w - (c^T + w\bar{\lambda}^T)x \geq w\left\|\eta P_0^T x\right\|_2, \tag{5.75'}$$

$$-\ b_i + \bar{a}_i^T x \geq \left\|\eta P_i^T x\right\|_2, \qquad i = 1,\ldots,N_c \tag{5.51'}$$

$$d - \bar{\lambda}^T x \geq \left\|\eta P_0^T\, x\right\|_2, \tag{5.52'}$$

$$-W_D + e^T x \geq 0, \tag{5.5}$$

$$Wc_i - x_i \geq 0, \qquad i = 1,\ldots,N_{iw}+N_{ew} \tag{5.6}$$

$$x_i \geq 0, \qquad i = 1,\ldots,N_{iw}+N_{ew} \tag{5.7}$$

5.4.7 The linearization approach in the robust multi-objective SW intrusion optimization problem

Recall that in the deterministic SW intrusion optimization problem, each decision (design) variable has only one response curve of SW intrusion increment. For a number of variables there are several response curves. The problem was to find in each curve j a point that has its fixed gradient (coefficient λ_j) for each corresponding variable x_j such that x_j is optimal. Then the sequential linearization approach will iteratively seek all fixed coefficients λ_j for all variables x_j for the objective (5.2) and constraint (5.4), and hence the optimal solution will be exactly achieved.

In the robust multi-objective SW intrusion optimization problem, graphically, the number of response curves of the SW intrusion increment for each design variable (a candidate well) are more than one. That number depends on the number of realizations. This means that the "mean curve" and a deviation from this curve to any curve among these response curves for each variable have to be computed. Therefore, similarly, the sequential linearization approach for the robust multi-objective SW intrusion optimization problem will iteratively seek the fixed "mean coefficients" $\overline{\lambda}_j$ and the fixed "deviation coefficients" P_0^T for the objective (5.50) and the constraint (5.52') such that with those fixed coefficients all x_j are optimal. The numerical results show that the problem will converge to the optimal solution faster if the value of ΔQ is gradually decreased when the iteration progresses (see Chapters 6 and 7).

5.4.8 Solution methodology

The steps for solving the robust multi-objective optimization problem will be as follows:

3. By using SGSIM generating ξ number of realizations of the uncertain parameter (in this case, the aquifer hydraulic conductivity field). Each realization, ξ, has different spatial values of the hydraulic conductivity, k.
30. By setting the stress equal to zero and a small perturb ΔQ the program will enter the first iteration of the sequential linearization approach.
31. Based on the two-fluid groundwater flow simulation software, SHARP, and with different realizations of hydraulic conductivity, the average intrusion length response matrix λ_{jk} and the groundwater head response matrices A_{jk} for each control point are computed.
32. The nominal vectors, $\overline{\lambda}, \overline{a}_i$, $i = 1, 2,\ldots, N_c$, and the perturbation matrices, ηP_0, ηP_i, $i = 1, 2,\ldots, N_c$, are computed.
33. Using the quadratic cone programming in SeDuMi, to solve each robust single objective problem subject to the same constraints. The optimal values of Z_1 and Z_2 are obtained as Z_{1opt} and Z_{2opt}, respectively.
34. Generating the non-inferior set in the objective space (Z_1, Z_2) by the constraint method in which the objective is Z_2 while Z_1 becomes the objective inequality constraint. The right-hand side of this objective constraint varies from Z_{1opt} to the corresponding value of Z_1 when $Z_2 = Z_{2opt}$.
35. If using the method of minimum distance from the ideal solution, the optimal values, Z_{1opt}, Z_{2opt} as computed in step 4 are taken as the ideal solution (utopia point). If using the method of prior assessments of weights then we need to articulate the preferences by assigning the value of weight ($w > 0$).
36. A multi-objective optimization problem, which finds the best-compromised solution by using either the method of minimum distance from the ideal solution or the method of prior assessments of weights, is solved using the quadratic cone programming in SeDuMi. The optimal solution to this problem, which is graphically checked by the generated non-inferior set, is taken as the stress input for the next iteration of the sequential linearization approach.
37. The sequential linearization approach will stop if the difference between the optimal solutions at two consecutive iterations less than or equal to the allowable error. Then the optimal solution of the last iteration will be the appropriate solution for implementation.

Figure 5.20 will show the whole procedures of the robust multi-objective optimization problem for SW intrusion management.

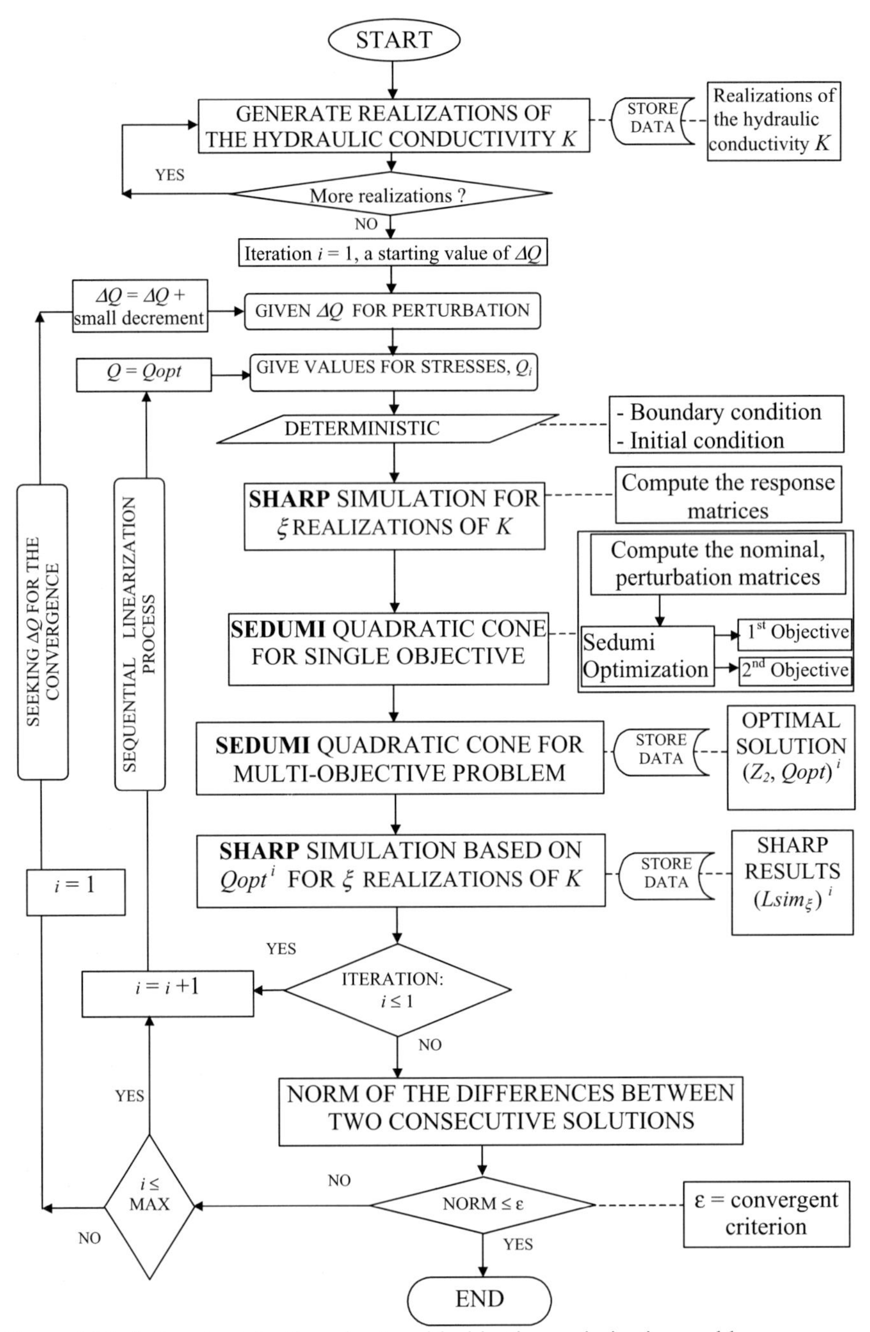

Figure 5.20: A flowchart for the robust multi-objective optimization problem.

Chapter 6

Hypothetical Example Results for the quasi-three-dimensional SHARP model simulation of a one-layered aquifer

6.1 Introduction

The hypothetical model that is based on the quasi-three-dimensional model as in chapter 3 consists of a rectangular area in the horizontal plane (see Figure 6.1). Its aquifer system is underlain by an upper layer with a low permeability, which is called an aquitard. The lower layer has higher permeability, which is called the aquifer. This aquifer is partly intruded by saltwater. The aquifer is considered as confined, having the same thickness in the cross sectional profile (see Figure 6.2).

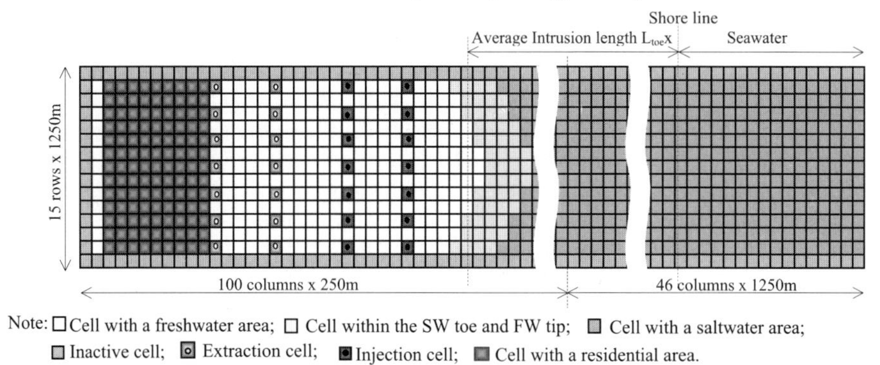

Figure 6.1: The horizontal plane view of the aquifer of the model.

In the horizontal plane, the two-dimensional grid system is simulated in which there are 15 rows of 1250m each in the y-direction. Grid spacing in the x-direction is 1250m except near the location of wells and the interface where it is reduced to 250m to improve the interface projection accuracy (see Essaid, 1990). Therefore the total number of columns in x-direction is 146. The SW intrusion length is taken as an average of the saltwater toe intrusion lengths of all rows in the horizontal plane. The salt/fresh interface is considered as an abrupt change of the groundwater density implying no existence of the mixing zone of brackish water.

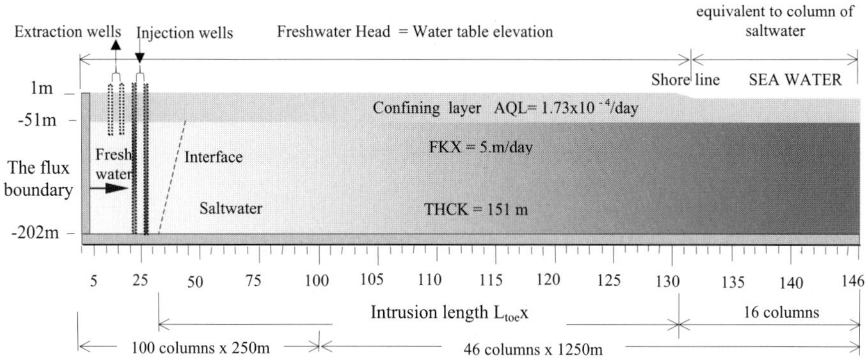

Figure 6.2: The cross sectional profile of the quasi-three-dimension model.

6.2 The deterministic SW intrusion optimization problem

The hydro-geological data and aquifer parameters in the deterministic problem will be as follows:

- The topmost one is the semi-pervious layer (aquitard), its leakance value is of approximately $1.73\ 10^{-4}$ /day ($\approx 2.0\ 10^{-9}$ /s).
- For simplicity, it is assumed that the permeable layer (main aquifer) is homogenous with its hydraulic conductivity of 5 m/day ($\approx 5.787\ 10^{-5}$ m/s) and its thickness of 151m.
- The bottom of the system is considered impermeable as it is a no-flow boundary.
- A constant flux boundary is put on the left boundary. Its value is assumed to be as large as possible, say 1800 m^3/h ($\approx 0.5\ m^3/s$).
- The partly penetrating extraction and fully penetrating injection wells are located at columns 12, 17 and 23, 28 respectively. The geometry of this problem is illustrated in Figures 6.1 and 6.2.

6.2.1 Objectives:

For the deterministic case of the saltwater intrusion management problem, the multi-objective linear programming under the set of N_w decision variables is formulated, which depends on the number of injection, extraction wells for which optimal values of pumping rates are desired. Here, it is necessary to observe the functioning of extraction and injection wells with respect to the distance of their wells from the interface toe by installing two parallel columns of extraction wells and two parallel columns of injection wells. The injection wells are installed near the interface toe to the freshwater zone, while the extraction wells are farther inland. The number of candidate extraction wells, N_{ew}, is 14 and so is the number of candidate injection wells, N_{iw}.

Here the management problem based on the sharp interface is emphasized by the movement of the saltwater with high salinity regardless of the brackish zone caused by the dynamic dispersion (see Stakelbeek, 1999).

Two objectives will be considered i.e.

- The minimization of the distance between the location of the interface toe and the shoreline at the end of the simulation time, when its steady state is achieved.

$$\min_{x}\ [Z_2 = \lambda^T x + L^o]$$

where L^o being the initial SW intrusion length (at non pumping condition) is not subject to minimization so it can be removed from the objective function to produce:

$$\min_{x}\ [Z_2 = \lambda^T x]$$

- The minimization of total costs due to extraction and injection strategies.

$$\min_{x}\ [Z_1 = c^T x]$$

where c^T is the vector of coefficients that represent the pumping and installation costs for extraction wells, c_{ext}, and injection wells, c_{inj}. The costs are transferred to the amount of monetary unit paid for every cubic meter of water in the problem. Here it is assumed that these costs are set constants. Thus this objective is a linear function with respect to x (the design vector of 28 decision variables of pumping rates).

Therefore the vector $c^T = [[c_{ext}]^T, [c_{inj}]^T]$ where $c_{ext} = 0.05$ MU/m^3 for each of 14 extraction wells and $c_{inj} = 0.1$ MU/m^3 for each of 14 injection wells. The costs for the injection are considered higher than for the extraction. This is because the treatment of river water must be done before the injection and the fully penetrating injection wells must cost more for installation.

Those two objectives are set such that they are not only conflicting (decreasing the intrusion length will necessarily require an increase of injection rate and then the operational cost) but also non-commensurate (different units in the objectives attributes) and hence the complexity involved in the solution process.

6.2.2 Constraints:

Surface water injection rate upper bounds

The maximum injection rate is only restricted by the capacity of the injection well that is much dependent on the well structure and the injection type (see Liu Peimin, et al., 1994). Here 0.25 m^3/s ($\approx$ 900 m^3/h) is set as the upper bound of all injection wells at one cell (equivalent to 10 wells/cell $\times$ 90m^{3}h^{-1}/well). Though the freshwater source (from the river) for the injection could be another limitation, here it is assumed that the river discharge in the flood season is sufficient to be unlimited for injecting.

$$x_j \le M_{inj} = 900, \qquad j = N_{ew} + 1, \ldots, N_{ew} + N_{iw}$$

Extraction-well capacity constraints

Pumping rates from potential extraction cells are limited to a maximum yield of 50m^3/h ($\approx$ 0.014 m^3/s).

$$x_j \le M_{ext} = 50, \qquad j = 1, \ldots, N_{ew}$$

Supplying demand constraints

The minimum water demand, W_d, for the region is of about 360 m^3/h (= 0.1 m^3/s). This is only for domestic use.

$$\sum_{j=1}^{N_{ew}} x_j \ge W_D = 360$$

Draw down constraints

The maximum draw down of piezometric head in the freshwater zone is not allowed to be greater than 10m because of the prevention of subsidence for the residential area. Hence $b = -10$. In the residential area there are 117 cells in total (see Figure 6.1) that are considered as the number of control points for the problem, $N_c = 117$. Thus there will be 117 draw down constraints under the forms of:

$$\sum_{j=1}^{N_{ew}+N_{iw}} A_{ij} x_j \ge b_i = -10, \qquad i = 1, \ldots, N_c = 117,$$

$$j = 1, \ldots, N_w = 28.$$

Saltwater toe location constraint

The maximum intrusion length is defined by the constraint of the increment of saltwater toe location. This is chosen so that the interface toe cannot reach either the well screen locations or the initial (at the beginning of pumping period) interface toe location during operation. This is called the first option of SW intrusion management. The second option for this constraint that will be selected is to extend the freshwater zone during operational process. This means that during the pumping period it is preferable to store more freshwater in the aquifer than to exploit. For the latter option, a distance of 55km landward from the shoreline (100 cells away from shoreline) is the upper bound of the intrusion length, L_{max}. The distance of the initial interface toe from the shoreline, L_0 is set about 55,286 m. Then the SW intrusion increment, $d = L_{max} - L_0$, is about –286m. For the first option it can be seen that $d = 0$.

$$\sum_{j=1}^{N_{ew}+N_{iw}} \lambda_j x_j \le d$$

Non-negative constraints

The non-negativity can be imposed on the pumping rates of the candidate wells by the non-negativity constraints as:

$$x_j \ge 0, \qquad j = 1,\ldots,N_{ew}+N_{iw}$$

This type of constraints guarantees that the optimal solution will always be positive. Consequently in the optimal solution vector, very small values (less than 10^{-3} m^3/h) for its components may occur that can be ignored and reported as zeros.

In total, the multi-objective SW intrusion management problem has the form:

$$\operatorname*{Min}_x \ [Z_1 = c^T x], \tag{6.1}$$

$$\operatorname*{Min}_x \ [Z_2 = \lambda^T x], \tag{6.2}$$

subject to

$$\sum_{j=1}^{N_{ew}+N_{iw}} A_{ij} x_j \ge b_i \ , \qquad i = 1,\ldots,N_c \tag{6.3}$$

$$\sum_{j=1}^{N_{ew}+N_{iw}} \lambda_j x_j \le d \ , \tag{6.4}$$

$$\sum_{j=1}^{N_{ew}} x_j \ge W_D \ , \tag{6.5}$$

$$x_j \le M_j, \begin{cases} M_j = M_{ext} & j = 1,\ldots,N_{ew} \\ M_j = M_{inj} & j = N_{ew}+1,\ldots,N_{ew}+N_{iw} \end{cases} \tag{6.6}$$

$$x_j \ge 0 \ , \qquad j = 1,\ldots,N_{ew}+N_{iw} \tag{6.7}$$

where $c^T = [0.05, 0.1]^T$, $b_i = -10$, $d = -286$, $W_D = 360$, $M_{ext} = 50$, $M_{inj} = 900$, λ_j and A_{ij} are computed by the SHARP and PERTURB computer codes.

6.2.3 The multi-objective optimal solution by the method of minimum distance from the ideal solution (L_2-metric)

The multi-objective optimization problem is rewritten as the same as in part 5.2:

$$\text{Min} \quad \delta, \tag{6.8}$$

s.t.

$$\left\| Z^* - Z(x) \right\|_2 \leq \delta, \tag{6.9}$$

$$\forall x \in \mathbf{F}_d, \tag{6.10}$$

where $\mathbf{F}_d$ is the feasible region in the decision space that is created by the linear constraints from (6.3),…,(6.7). $\left\| Z^* - Z(x) \right\|_2 \leq \delta$ is the non-linear-objective constraint. Now the new objective function is linearly determined on the set of design variable (δ, x).

To solve the problem the computer code has been programmed by the author so that the SHARP executable code and SeDuMi optimization solver can be combined in one file of the Matlab language. The following are the results of this problem:

Results

1. The optimal solution $x = \boldsymbol{Q}_{opt}$, at all iterations is shown in Table 6.1. The value of ΔQ equal to 10 m^3/h is initially applied to the first iteration and it gradually decreases by steps of 0.2 m^3/h for subsequent iterations. The optimal solution converges at the third iteration.

2. The optimal (trade-off) values of Z_1 and Z_2 and the SW intrusion increment, *Lsim*, which are shown in Table 6.2, are based on the optimal solution at different iterations. The trade-off values of (Z_1, Z_2) in the non-inferior set at the first and last iterations are depicted in Figures 6.3 and 6.4. The final (third) optimal value of Z_2 is computed as −2669.870 m which is relatively close to −2615.987 m being the simulated saltwater intrusion increment. The relative difference is about 2%. The negative sign of the saltwater intrusion increment or the optimal value Z_2 indicates the decrement of the saltwater intrusion from the initial interface toe (at non-pumping condition) by the distance equal to that absolute value.

The two Figures 6.3 and 6.4 show a considerable change of shape of the two non-inferior sets. Hence, there is a big difference between the optimal values at two iterations. For example, with L_2-metric solution, the optimal values for the first iteration are (630.25, −567.75) and the optimal values for the last (3rd) iteration are (648, −2669.87). This proves the importance of requiring the linearization approach for such a non-linear problem in optimization. These differences are not only because of the variation of the optimal solution to different iterations but also because the great changes of the coefficients of the saltwater intrusion response matrices between the two-consecutive iterations exist until the optimal solution converges.

Note that the non-inferior sets drawn in Figures 6.3 and 6.4 are created by the constraint method with the response matrices $[\Delta H_i / \Delta Q_j]$ and $[\Delta L / \Delta Q_j]$. These are based on the optimal solution from the L_2-metric problem. The optimal values of the L_1-metric and L_{inf}-metric computed are based on the same response matrices generated from the L_2-metric problem. Those values of the two problems (L_1-metric and L_{inf}-metric) are just for showing the narrow range of choice in the non-inferior set for the L_α-metric problems when α varies from 1 to infinity. This procedure of generating the non-inferior set and computing the optimal values of the L_α-metric is applicable for all the L_α-metric problems in this chapter.

3. The optimal solution $x = \boldsymbol{Q}_{opt}$, which is shown in Table 6.3, is obtained at the third iteration ($k = 3$). This solution is found to be symmetric for both the extraction rates and the injection rates. These values of $\boldsymbol{Q}_{opt}$ can be referred to Figure 6.7 which is the output file of SHARP for mapping the extent of SW intrusion based on the last-iteration optimal solution.

4. The SW intrusion interface is simulated by SHARP computer code. The maps in Figures 6.5, 6.6, 6.7, which show the extent of saltwater intrusion, are the outputs of the simulation model based on the optimal solutions of the corresponding iterations. The map of the third iteration that is based on the converged optimal solution is the final simulated result map for the problem. All maps show the symmetry of both the interface tip and toe. The locations of the interfaces in the plan view are quite regular from one row to another.

Discussion

Not only for this case, the symmetry of the interface tip and toe exists since homogeneity is assumed for the distributed transmissivities of the aquifer, and the optimal solution obtained from the L_2-metric multi-objective problem in the deterministic case is also symmetric.

The symmetry of the optimal solution can be explained by the fact that the multi-objective problem is subject to the number of constraints that are symmetric or equalized for all control points of the modeled area. Firstly, the saltwater intrusion length constraint is averaged for all rows of the modeled area; secondly, the freshwater head constraints are equalized for all control points; and thirdly, the well-capacity constraints are set equally for all extraction wells as well as for all injection wells. These mentioned constraints allow the calculation of the optimal solution that will be symmetric, as long as the response matrices with respect to extraction or injection are symmetric. In the deterministic problems, the candidate wells (for extraction or injection) are arranged in a symmetric way so that their response matrices will also be symmetric if the transmissivities are constant for the whole aquifer.

The optimal solution obtained in this L_2-metric multi-objective problem is optimized for the seven extraction wells in the same column 12, one extraction well in the middle row of column 17, and the seven injection wells in the same column 28. Such a well allocation will cause the regular shape of interfaces in the plan view.

The slope of the non-inferior set ($\mathbf{N}_o$) which is defined as a ratio of the saltwater intrusion decrement (ΔZ_2) to the injection cost increment (ΔZ_1) is called the marginal rate of transformation (MRT). In Figure 6.4 we can split up the $\mathbf{N}_o$ in two sets such that for points (Z_1, Z_2) in the first set the MRT is always bigger than for points in the second set. It reflects the fact that application of higher costs for injection is after the point (648, –2669.86) less effective.

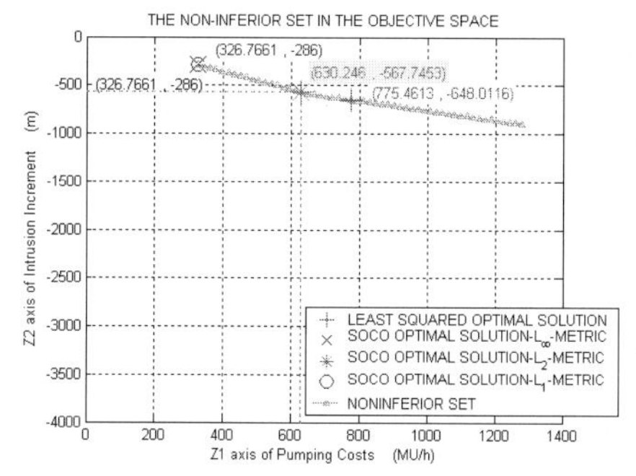

Figure 6.3: At iteration 1.

Figure 6.4: At iteration 3.

Table 6.1: The results of the L_2-metric problem under deterministic case.

	Iteration number, k		
	1	2	3
ΔQ	10.000	9.800	9.600
Z_1	630.246	648.000	648.000
Z_2	-567.745	-2698.170	-2669.870
$Lsim$	-2567.133	-2615.987	-2615.987
$\|(Q_i^{\ k} - Q_i^{\ k-1})\|$	100000.000	177.540	0.000
Q_1	50.000	50.000	50.000
:	50.000	50.000	50.000
:	50.000	50.000	50.000
:	50.000	50.000	50.000
:	50.000	50.000	50.000
:	50.000	50.000	50.000
:	50.000	50.000	50.000
:	0.000	0.000	0.000
:	0.000	0.000	0.000
:	0.000	0.000	0.000
:	10.000	10.000	10.000
:	0.000	0.000	0.000
:	0.000	0.000	0.000
:	0.000	0.000	0.000
:	0.000	0.000	0.000
:	0.000	0.000	0.000
:	0.000	0.000	0.000
:	0.000	0.000	0.000
:	0.000	0.000	0.000
:	0.000	0.000	0.000
:	0.000	0.000	0.000
:	900.000	900.000	900.000
:	900.000	900.000	900.000
:	900.000	900.000	900.000
:	722.460	900.000	900.000
:	900.000	900.000	900.000
:	900.000	900.000	900.000
Q_{28}	900.000	900.000	900.000

Table 6.2: The optimal values and the *Lsim* at all different iterations (L_2-metric).

Iteration	ΔQ (m^3/h)	Z_1 (MU/h)	Z_2 (m)	*Lsim* (m)	$\|(Q_i^k - Q_i^{k-1})\|$
1	10.0	630.246	-567.745	-2567.133	100000.000
2	9.8	648.000	-2698.170	-2615.987	177.540
3	9.6	648.000	-2669.870	-2615.987	0.000

Table 6.3: The optimal solution at the last iteration (L_2-metric)

Q_{opt} (m^3/h)	Column 12	Column 17	Column 23	Column 28
Row 14	50	0	0	900
Row 12	50	0	0	900
Row 10	50	0	0	900
Row 8	50	10	0	900
Row 6	50	0	0	900
Row 4	50	0	0	900
Row 2	50	0	0	900

```
MAP OF EXTENT OF INTRUSION (F-FRESHWATER, M-FRESH AND SALTWATER, S-SALTWATER)
                   10        20        30        40        50        60         146
          1234567890123456789012345678901234567890134567890123456789012345  12456
   15     ................................................................  .....
   14     .FRRRRRRRRRREFFFFEFFFFFIFFFFIMMMMMMMMMSSSSSSSSSSSSSSSSSSSSSSSSSSS  SSSSS
   13     .FRRRRRRRRRFFFFFFFFFFFFFFFFFMMMMMMMMMSSSSSSSSSSSSSSSSSSSSSSSSSSS  SSSSS
   12     .FRRRRRRRRREFFFFEFFFFFIFFFFIMMMMMMMMMSSSSSSSSSSSSSSSSSSSSSSSSSSS  SSSSS
   11     .FRRRRRRRRRFFFFFFFFFFFFFFFFFMMMMMMMMMSSSSSSSSSSSSSSSSSSSSSSSSSSS  SSSSS
   10     .FRRRRRRRRREFFFFEFFFFFIFFFFIMMMMMMMMMSSSSSSSSSSSSSSSSSSSSSSSSSSS  SSSSS
    9     .FRRRRRRRRRFFFFFFFFFFFFFFFFFMMMMMMMMMSSSSSSSSSSSSSSSSSSSSSSSSSSS  SSSSS
    8     .FRRRRRRRRREFFFFEFFFFFIFFFFIMMMMMMMMMSSSSSSSSSSSSSSSSSSSSSSSSSSS  SSSSS
    7     .FRRRRRRRRRFFFFFFFFFFFFFFFFFMMMMMMMMMSSSSSSSSSSSSSSSSSSSSSSSSSSS  SSSSSS
    6     .FRRRRRRRRREFFFFEFFFFFIFFFFIMMMMMMMMMSSSSSSSSSSSSSSSSSSSSSSSSSSS  SSSSSS
    5     .FRRRRRRRRRFFFFFFFFFFFFFFFFFMMMMMMMMMSSSSSSSSSSSSSSSSSSSSSSSSSSS  SSSSSS
    4     .FRRRRRRRRREFFFFEFFFFFIFFFFIMMMMMMMMMSSSSSSSSSSSSSSSSSSSSSSSSSSS  SSSSSS
    3     .FRRRRRRRRRFFFFFFFFFFFFFFFFFMMMMMMMMMSSSSSSSSSSSSSSSSSSSSSSSSSSS  SSSSSS
    2     .FRRRRRRRRREFFFFEFFFFFIFFFFIMMMMMMMMMSSSSSSSSSSSSSSSSSSSSSSSSSSS  SSSSSS
    1     ................................................................  .....
          1234567890123456789012345678901234567890134567890123456789012345  23456
                   10        20        30        40        50        60         146
(NOTE: R-RESIDENTIAL AREA. E- CANDIDATE EXTRACTION. I- CANDIDATE INJECTION)
```

Figure 6.5: The SW toe and FW tip at the non-pumping condition.

```
MAP OF EXTENT OF INTRUSION (F-FRESHWATER, M-FRESH AND SALTWATER, S-SALTWATER)
                   10        20        30        40        50        60         146
          1234567890123456789012345678901234567890134567890123456789012345  12456
   15     ................................................................  .....
   14     .FRRRRRRRRREFFFFEFFFFFIFFFFIFFFFFFFFFFMMMMMMMMMMSSSSSSSSSSSSSSSS  SSSSS
   13     .FRRRRRRRRRFFFFFFFFFFFFFFFFFFFFFFFFFFFMMMMMMMMMMSSSSSSSSSSSSSSSS  SSSSS
   12     .FRRRRRRRRREFFFFEFFFFFIFFFFIFFFFFFFFFFMMMMMMMMMMSSSSSSSSSSSSSSSS  SSSS
   11     .FRRRRRRRRRFFFFFFFFFFFFFFFFFFFFFFFFFFFMMMMMMMMMMSSSSSSSSSSSSSSSS  SSSSS
   10     .FRRRRRRRRREFFFFEFFFFFIFFFFIFFFFFFFFFFMMMMMMMMMMSSSSSSSSSSSSSSSS  SSSSS
    9     .FRRRRRRRRRFFFFFFFFFFFFFFFFFFFFFFFFFFFMMMMMMMMMMSSSSSSSSSSSSSSSS  SSSSS
    8     .FRRRRRRRRREFFFFEFFFFFIFFFFIFFFFFFFFFFMMMMMMMMMMSSSSSSSSSSSSSSSS  SSSSS
    7     .FRRRRRRRRRFFFFFFFFFFFFFFFFFFFFFFFFFFFMMMMMMMMMMSSSSSSSSSSSSSSSS  SSSSSS
    6     .FRRRRRRRRREFFFFEFFFFFIFFFFIFFFFFFFFFFMMMMMMMMMMSSSSSSSSSSSSSSSS  SSSSSS
    5     .FRRRRRRRRRFFFFFFFFFFFFFFFFFFFFFFFFFFFMMMMMMMMMMSSSSSSSSSSSSSSSS  SSSSSS
    4     .FRRRRRRRRREFFFFEFFFFFIFFFFIFFFFFFFFFFMMMMMMMMMMSSSSSSSSSSSSSSSS  SSSSSS
    3     .FRRRRRRRRRFFFFFFFFFFFFFFFFFFFFFFFFFFFMMMMMMMMMMSSSSSSSSSSSSSSSS  SSSSSS
    2     .FRRRRRRRRREFFFFEFFFFFIFFFFIFFFFFFFFFFMMMMMMMMMMSSSSSSSSSSSSSSSS  SSSSSS
    1     ................................................................  .....
          1234567890123456789012345678901234567890134567890123456789012345  23456
                   10        20        30        40        50        60         146
NOTE: R-RESIDENTIAL AREA, E-CANDIDATE EXTRACTION, I- CANDIDATE INJECTION, E-OPTIMAL EXTRACTION, I-OPTIMAL INJECTION
```

Figure 6.6: The SW toe and FW tip after the first iteration.

```
MAP OF EXTENT OF INTRUSION (F-FRESHWATER, M-FRESH AND SALTWATER, S-SALTWATER)
                 10        20        30        40        50        60         146
         1234567890123456789012345678901234567890134567890123456789012345  12456
   15    ................................................................  .....
   14    .FRRRRRRRRRE FFFFEFFFFFIFFFFIFFFFFFFFFFFFMMMMMMMMMMMSSSSSSSSSSSSSSS  SSSSS
   13    .FRRRRRRRRRFFFFFFFFFFFFFFFFFFFFFFFFFFFFFFMMMMMMMMMMMSSSSSSSSSSSSSSS  SSSSS
   12    .FRRRRRRRRRE FFFFEFFFFFIFFFFIFFFFFFFFFFFFMMMMMMMMMMMSSSSSSSSSSSSSSS  SSSSS
   11    .FRRRRRRRRRFFFFFFFFFFFFFFFFFFFFFFFFFFFFFMMMMMMMMMMMMSSSSSSSSSSSSSSS  SSSSS
   10    .FRRRRRRRRRE FFFFEFFFFFIFFFFIFFFFFFFFFFFMMMMMMMMMMMMSSSSSSSSSSSSSSS  SSSSS
    9    .FRRRRRRRRRFFFFFFFFFFFFFFFFFFFFFFFFFFFFFMMMMMMMMMMMMSSSSSSSSSSSSSS   SSSSS
    8    .FRRRRRRRRRE FFFFEFFFFFIFFFFIFFFFFFFFFFFMMMMMMMMMMMMSSSSSSSSSSSSSS   SSSSS
    7    .FRRRRRRRRRFFFFFFFFFFFFFFFFFFFFFFFFFFFFFMMMMMMMMMMMMSSSSSSSSSSSSSS  SSSSSS
    6    .FRRRRRRRRRE FFFFEFFFFFIFFFFIFFFFFFFFFFFMMMMMMMMMMMMSSSSSSSSSSSSSS  SSSSSS
    5    .FRRRRRRRRRFFFFFFFFFFFFFFFFFFFFFFFFFFFFFMMMMMMMMMMMMSSSSSSSSSSSSSS  SSSSSS
    4    .FRRRRRRRRRE FFFFEFFFFFIFFFFIFFFFFFFFFFFFMMMMMMMMMMMSSSSSSSSSSSSSS  SSSSSS
    3    .FRRRRRRRRRFFFFFFFFFFFFFFFFFFFFFFFFFFFFFFMMMMMMMMMMMSSSSSSSSSSSSSS  SSSSSS
    2    .FRRRRRRRRRE FFFFEFFFFFIFFFFIFFFFFFFFFFFFMMMMMMMMMMMSSSSSSSSSSSSSS  SSSSSS
    1    .......................................................................
         1234567890123456789012345678901234567890134567890123456789012345  23456
                 10        20        30        40        50        60         146
NOTE: R-RESIDENTIAL AREA, E-CANDIDATE EXTRACTION, I- CANDIDATE INJECTION, E-OPTIMAL EXTRACTION, I-OPTIMAL INJECTION
```

Figure 6.7: The SW toe and FW tip after the third iteration.

6.2.4 The multi-objective optimal solution by the method of prior assessments of weights (weighted problem)

Here another optimal solution is illustrated by using the method of prior assessments of weights for the deterministic multi-objective problem of SW intrusion management. With $w = 0.1, 0.6, 2.2 > 0$, different optimal solutions for the multi-objective problem from (6.1) to (6.7) can be obtained. These are written in the general form:

$$\text{Min} \qquad c^T x + w\lambda^T x \tag{5.35}$$

s.t.

$$x \in \mathbf{F}_\mathrm{d} \tag{5.11'}$$

where $\mathbf{F}_\mathrm{d}$ is the feasible region in the decision space. $\mathbf{F}_\mathrm{d}$ is determined by the constraints from (6.3) to (6.7); w is the value of weight, $w > 0$.

Results

1. The optimal solution $x = \boldsymbol{Q}_{opt}$

The optimal solutions of all iterations for the weighted problem ($w = 0.6$) are shown in Table 6.4. The optimal solution in this problem is converged in the third iteration. The optimal solutions at the three iterations are used in the input file for SHARP simulation program. The results are correspondingly shown in Figures 6.11 and 6.12 for the first and last iterations.

2. The optimal (trade-off) values of Z_1 and Z_2 and the SW intrusion increment, *Lsim*, which are computed and shown in Table 6.5, are based on the optimal solution at different iterations. The optimal value (Z_2) obtained at the third iteration is −1477.674 m. If compared to the simulated saltwater intrusion increment which is −1486.893 m, the relative difference is quite small (about 0.6%). The trade-off values of (Z_1, Z_2) in the non-inferior set at the first and last iteration are depicted in Figures 6.8 and 6.9.

Note that the non-inferior sets drawn in Figures 6.8 and 6.9 are created by the constraint method with the response matrices [$\Delta H_i / \Delta Q_j$] and [$\Delta L / \Delta Q_j$]. These are based on the optimal solution from the $w = 0.6$ weighted problem. The optimal values with the weights w = 0.1 and w = 2.2 are based on the same response matrices generated from the w = 0.6 weighted problem. Those values of the two problems (with w = 0.1 and w = 2.2) are just for showing the full range of choice in the non-inferior set for the weighted problem. Such a procedure of generating the non-inferior set and computing the optimal values is applicable for all the weighted problems in this chapter.

By comparison of the two Figures 6.8 and 6.9, it is possible to also see a big difference between the optimal values at two iterations, e.g. with $w = 0.6$, the optimal values for the first iteration are (326.76, −286) and the optimal values for the last iteration are (378, −1477.67).

For the non-inferior set in Figure 6.9, the optimal values (Z_1 = 378 MU/h, Z_2 = −1477.67 m) in the weighted problem with $w = 0.6$ are more in favour of minimizing Z_1 if compared to the optimal values (498.2 MU/h, -1589.03 m) in L_2-metric method.

3. The optimal solution $x = \boldsymbol{Q}_{opt}$, which is shown in Table 6.6, is obtained at the third iteration ($k = 3$). The optimal solution in the weighted problem also shows the symmetry for both the extraction and injection rates. The reader can refer these values of $\boldsymbol{Q}_{opt}$ to Figure 6.12 which is the output file of SHARP for mapping the extent of SW intrusion based on the last-iteration optimal solution vector.

4. The SW intrusion interface is simulated by SHARP computer code. Figure 6.10 is the output file of the SHARP simulation program when running with $Q_j = 0$. This result is used for computing the SW intrusion length, L^o, at the non-pumping condition for the optimization program. Based on the optimal solutions of the iterations, the simulated maps of saltwater toe and freshwater tip at these iterations are also symmetric. The similar reason for the symmetry of the interface tip and toe is discussed as in the L_2-metric problem.

The average value of the simulated saltwater intrusion increments (−1486.89 m) in this problem is greater than the one (−2615.98 m) in the L_2-metric problem.

The locations of the interfaces in the plan view are quite different from one row to another. The interface tip and toe of rows near the left and right boundaries move farther seaward than the ones in the middle rows. This can be explained by the functioning of four injection wells in the column 28 ($Q_{injection}$ = 900 m^3/h). The two injection wells with the coordinates (column, row) of (28,12) and (28,14) are near the left and the other two injection wells of (28,2) and (28,4) are near the right boundaries.

Discussion

In this problem, the weight value, w = 0.6, which is quite small, shows that the preferences of the decision maker are more in favor of minimization of the Z_1 objective if compared to the L_2-metric problem. Therefore, in terms of the injection, the obtained optimal solution is allocated for only four injection wells in the same column 28. That will result in the non-regular shape of the interfaces in the plan view and the greater average value ($Lsim$ = −1486.893 m) of the simulated saltwater intrusion increments. The latter can be checked by the optimal value Z_2 = −1477.674 m with the small difference of 0.6%.

Also with this problem, the trade-off of the optimal values (Z_1, Z_2) between two methods can be seen. In the weighted problem, Z_1 = 378 MU/h is smaller than Z_1 =

648 MU/h in the L_2-metric problem, whereas, Z_2 = −1477.67 m of the weighted problem is greater than Z_2 = −2669.87 m in the L_2-metric problem, inversely.

The allocation of those four injection wells is optimized based on the smallest response coefficients (negative values) of these injection well variables in the saltwater intrusion length response matrix (λ_j).

The allocation of the seven extraction wells ($Q_{extract}$ = 50 m^3/h) located in column 12 is optimized based on the smallest response coefficients (positive values) of the extraction well variables in λ_j. The additional extraction well ($Q_{extract}$ = 10 m^3/h) in column 17 is optimized based on the remaining of 10 m^3/h of the minimal water demand (w_d = 360 m^3/h) constraint and the smallest response coefficient at the middle row (row 8) in column 17.

In Figure 6.9, the lower part of the non-inferior set with which the MRT is smaller can be found also for this weighted problem.

Table 6.4: The results of the weighted problem, w = 0.6, under deterministic case.

	Iteration number, k		
	1	2	3
ΔQ	10.000	9.800	9.600
Z_1	326.766	378.000	378.000
Z_2	-286.000	-1491.699	-1477.674
Lsim	-1287.928	-1486.893	-1486.893
$\lVert(Q_i^{\,k} - Q_i^{\,k-1})\rVert$	100000.000	512.339	0.000
Q_1	50.000	50.000	50.000
:	50.000	50.000	50.000
:	50.000	50.000	50.000
:	50.000	50.000	50.000
:	50.000	50.000	50.000
:	50.000	50.000	50.000
:	50.000	50.000	50.000
:	0.000	0.000	0.000
:	0.000	0.000	0.000
:	0.000	0.000	0.000
:	10.000	10.000	10.000
:	0.000	0.000	0.000
:	0.000	0.000	0.000
:	0.000	0.000	0.000
:	0.000	0.000	0.000
:	0.000	0.000	0.000
:	0.000	0.000	0.000
:	0.000	0.000	0.000
:	0.000	0.000	0.000
:	0.000	0.000	0.000
:	0.000	0.000	0.000
:	900.000	900.000	900.000
:	900.000	900.000	900.000
:	0.000	0.000	0.000
:	0.000	0.000	0.000
:	0.000	0.000	0.000
:	387.661	900.000	900.000
Q_{28}	900.000	900.000	900.000

Table 6.5: The optimal values and the *Lsim* at all different iterations ($w = 0.6$).

Iteration	ΔQ (m^3/h)	Z_1 (MU/h)	Z_2 (m)	*Lsim* (m)	$\|(Q_i^{\,k} - Q_i^{\,k-1})\|$
1	10.0	326.766	-286.000	-1287.928	100000.000
2	9.8	378.000	-1491.699	-1486.893	512.339
3	9.6	378.000	-1477.674	-1486.893	0.000

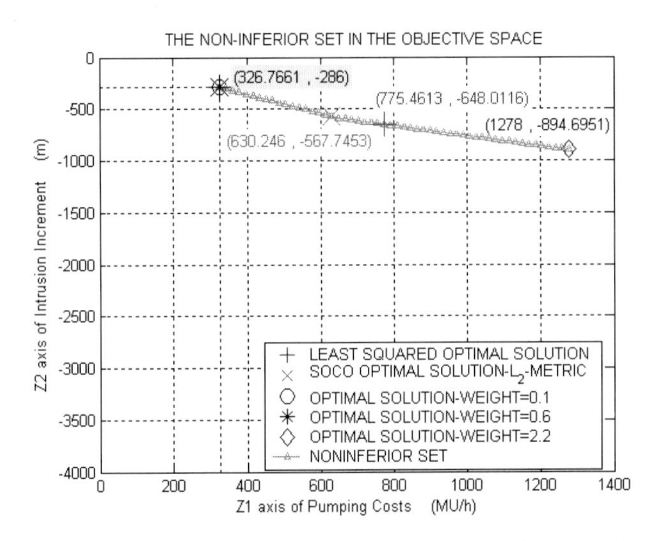

Figure 6.8: The non-inferior set in the objective space at the first iteration.

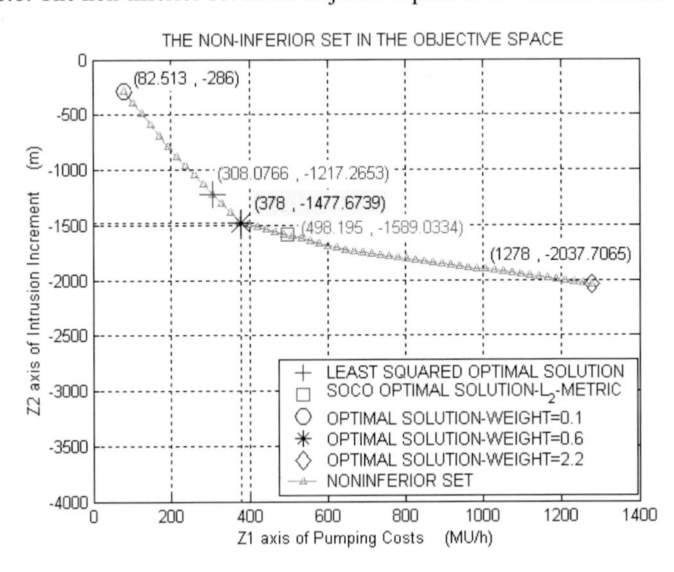

Figure 6.9: The non-inferior set in the objective space at the third iteration.

Table 6.6: The optimal solution at the last iteration ($w = 0.6$)

Q_{opt} (m^3/h)	Column 12	Column 17	Column 23	Column 28
Row 14	50.000	0.000	0.000	900.000
Row 12	50.000	0.000	0.000	900.000
Row 10	50.000	0.000	0.000	0.000
Row 8	50.000	10.000	0.000	0.000
Row 6	50.000	0.000	0.000	0.000
Row 4	50.000	0.000	0.000	900.000
Row 2	50.000	0.000	0.000	900.000

```
MAP OF EXTENT OF INTRUSION (F-FRESHWATER, M-FRESH AND SALTWATER, S-SALTWATER)
                 10        20        30        40        50        60         146
        12345678901234567890123456789012345678901345678901234567890123456789012345  12456
  15    .................................................................  .....
  14    .FRRRRRRRRREFFFFEFFFFFIFFFFIMMMMMMMMMSSSSSSSSSSSSSSSSSSSSSSSSSSSS  SSSSS
  13    .FRRRRRRRRRFFFFFFFFFFFFFFFFFMMMMMMMMMSSSSSSSSSSSSSSSSSSSSSSSSSSSS  SSSSS
  12    .FRRRRRRRRREFFFFEFFFFFIFFFFIMMMMMMMMMSSSSSSSSSSSSSSSSSSSSSSSSSSSS  SSSSS
  11    .FRRRRRRRRRFFFFFFFFFFFFFFFFFMMMMMMMMMSSSSSSSSSSSSSSSSSSSSSSSSSSSS  SSSSS
  10    .FRRRRRRRRREFFFFEFFFFFIFFFFIMMMMMMMMMSSSSSSSSSSSSSSSSSSSSSSSSSSSS  SSSSS
   9    .FRRRRRRRRRFFFFFFFFFFFFFFFFFMMMMMMMMMSSSSSSSSSSSSSSSSSSSSSSSSSSSS  SSSSS
   8    .FRRRRRRRRREFFFFEFFFFFIFFFFIMMMMMMMMMSSSSSSSSSSSSSSSSSSSSSSSSSSS   SSSSS
   7    .FRRRRRRRRRFFFFFFFFFFFFFFFFFMMMMMMMMMSSSSSSSSSSSSSSSSSSSSSSSSSSS  SSSSSS
   6    .FRRRRRRRRREFFFFEFFFFFIFFFFIMMMMMMMMMSSSSSSSSSSSSSSSSSSSSSSSSSSS  SSSSSS
   5    .FRRRRRRRRRFFFFFFFFFFFFFFFFFMMMMMMMMMSSSSSSSSSSSSSSSSSSSSSSSSSS  SSSSSS
   4    .FRRRRRRRRREFFFFEFFFFFIFFFFIMMMMMMMMMSSSSSSSSSSSSSSSSSSSSSSSSSS  SSSSSS
   3    .FRRRRRRRRRFFFFFFFFFFFFFFFFFMMMMMMMMMSSSSSSSSSSSSSSSSSSSSSSSSSS  SSSSSS
   2    .FRRRRRRRRREFFFFEFFFFFIFFFFIMMMMMMMMMSSSSSSSSSSSSSSSSSSSSSSSSSS  SSSSSS
   1    .................................................................  .....
        12345678901234567890123456789012345678901345678901234567890123456789012345  23456
                 10        20        30        40        50        60         146
(NOTE: R-RESIDENTIAL AREA. E- CANDIDATE EXTRACTION. I- CANDIDATE INJECTION)
```

Figure 6.10: The SW toe and FW tip at the non-pumping condition.

```
MAP OF EXTENT OF INTRUSION (F-FRESHWATER, M-FRESH AND SALTWATER, S-SALTWATER)
                 10        20        30        40        50        60         146
        12345678901234567890123456789012345678901345678901234567890123456789012345  12456
  15    .................................................................  .....
  14    .FRRRRRRRRREFFFFEFFFFFIFFFFIFFFFFFFFFFMMMMMMMMMMMSSSSSSSSSSSSSSSS  SSSSS
  13    .FRRRRRRRRRFFFFFFFFFFFFFFFFFFFFFFFFFMMMMMMMMMMMMSSSSSSSSSSSSSSSSS  SSSSS
  12    .FRRRRRRRRREFFFFEFFFFFIFFFFIFFFFFFFMMMMMMMMMMMMSSSSSSSSSSSSSSSSSS  SSSSS
  11    .FRRRRRRRRRFFFFFFFFFFFFFFFFFFFFFMMMMMMMMMMMSSSSSSSSSSSSSSSSSSSSSS  SSSSS
  10    .FRRRRRRRRREFFFFEFFFFFIFFFFIFMMMMMMMMMMMMMSSSSSSSSSSSSSSSSSSSSSSS  SSSSS
   9    .FRRRRRRRRRFFFFFFFFFFFFFFFFFMMMMMMMMMMMMSSSSSSSSSSSSSSSSSSSSSSSSS  SSSSS
   8    .FRRRRRRRRREFFFFEFFFFFIFFFFIMMMMMMMMMMMSSSSSSSSSSSSSSSSSSSSSSSSSS  SSSSSS
   7    .FRRRRRRRRRFFFFFFFFFFFFFFFFFMMMMMMMMMMMMMMMSSSSSSSSSSSSSSSSSSSSSS  SSSSSS
   6    .FRRRRRRRRREFFFFEFFFFFIFFFFIFFFMMMMMMMMMMMMMSSSSSSSSSSSSSSSSSSSSS  SSSSSS
   5    .FRRRRRRRRRFFFFFFFFFFFFFFFFFFFFFFFFMMMMMMMMMMMSSSSSSSSSSSSSSSSSSS  SSSSSS
   4    .FRRRRRRRRREFFFFEFFFFFIFFFFIFFFFFFFFFMMMMMMMMMMSSSSSSSSSSSSSSSSSS  SSSSSS
   3    .FRRRRRRRRRFFFFFFFFFFFFFFFFFFFFFFFFFFFMMMMMMMMMMSSSSSSSSSSSSSSSSS  SSSSSS
   2    .FRRRRRRRRREFFFFEFFFFFIFFFFIFFFFFFFFFFFMMMMMMMMMMSSSSSSSSSSSSSSSS  SSSSS
   1    .......................................................................
        12345678901234567890123456789012345678901345678901234567890123456789012345  23456
                 10        20        30        40        50        60         146
NOTE: R-RESIDENTIAL AREA, E-CANDIDATE EXTRACTION, I- CANDIDATE INJECTION, E-OPTIMAL EXTRACTION, I-OPTIMAL INJECTION
```

Figure 6.11: The SW toe and FW tip after the first iteration.

```
MAP OF EXTENT OF INTRUSION (F-FRESHWATER, M-FRESH AND SALTWATER, S-SALTWATER)
                    10        20        30        40       50        60          146
           1234567890123456789012345678901234567890134567890123456789012345   12456
   15      ................................................................   .....
   14      .FRRRRRRRRRERFFFFEFFFFFIFFFFIFFFFFFFFFFFMMMMMMMMMMSSSSSSSSSSSSSSSS  SSSSS
   13      .FRRRRRRRRRFFFFFFFFFFFFFFFFFFFFFFFFFFFFFMMMMMMMMMMSSSSSSSSSSSSSSSSS SSSSS
   12      .FRRRRRRRRREFFFFEFFFFFIFFFFIFFFFFFFFFMMMMMMMMMMSSSSSSSSSSSSSSSSSSSS  SSSS
   11      .FRRRRRRRRRFFFFFFFFFFFFFFFFFFFFFFFFFMMMMMMMMMMMMSSSSSSSSSSSSSSSSSSSS SSSSS
   10      .FRRRRRRRRREFFFFEFFFFFIFFFFIFFFMMMMMMMMMMMMMMSSSSSSSSSSSSSSSSSSSSSSS SSSSS
    9      .FRRRRRRRRRFFFFFFFFFFFFFFFFFFMMMMMMMMMMMMMMSSSSSSSSSSSSSSSSSSSSSSSSS SSSSS
    8      .FRRRRRRRRREFFFFEFFFFFIFFFFIMMMMMMMMMMMMSSSSSSSSSSSSSSSSSSSSSSSSSSS SSSSSS
    7      .FRRRRRRRRRFFFFFFFFFFFFFFFFFFMMMMMMMMMMMMMMSSSSSSSSSSSSSSSSSSSSSSSS SSSSSS
    6      .FRRRRRRRRREFFFFEFFFFFIFFFFIFFFMMMMMMMMMMMMMMSSSSSSSSSSSSSSSSSSSSSS SSSSSS
    5      .FRRRRRRRRRFFFFFFFFFFFFFFFFFFFFFFFFFMMMMMMMMMMMMSSSSSSSSSSSSSSSSSSS SSSSSS
    4      .FRRRRRRRRREFFFFEFFFFFIFFFFIFFFFFFFFFMMMMMMMMMMSSSSSSSSSSSSSSSSSSS  SSSSSS
    3      .FRRRRRRRRRFFFFFFFFFFFFFFFFFFFFFFFFFFFFMMMMMMMMMMSSSSSSSSSSSSSSSSS  SSSSSS
    2      .FRRRRRRRRREFFFFEFFFFFIFFFFIFFFFFFFFFFFMMMMMMMMMMSSSSSSSSSSSSSSSS   SSSSSS
    1      ................................................................   ......
           1234567890123456789012345678901234567890134567890123456789012345   23456
                    10        20        30        40       50        60          146
NOTE: R-RESIDENTIAL AREA, E-CANDIDATE EXTRACTION, I- CANDIDATE INJECTION, E-OPTIMAL EXTRACTION, I-OPTIMAL INJECTION
```

Figure 6.12: The SW toe and FW tip after the third iteration.

6.3 The robust multi-objective problem for the SW intrusion management

In this problem the uncertainty of the hydraulic conductivity value is taken into account. Since the simulation model results depend on the random values of the hydraulic conductivity, thus the response coefficients of the objective and the constraints in the optimization problem now vary in a field of uncertainty, which is treated as ellipsoidal uncertainty sets. These are described in Chapter 5 (part 5.4). Therefore the optimization problem now becomes the stochastic optimization problem. The differences between the objectives and constraints of this problem and those of the deterministic optimization problem are that in the objective (6.2) and the constraints (6.3) and (6.4), are that, instead, λ and A are replaced by the nominal values, $\bar{\lambda}$, $\bar{a}_i$, and the perturbation matrices, P_0, P_i, in the objective below:

$$\text{Min} \quad \bar{\lambda}^T x + \left\| P_0^T x \right\|_2$$

and constraints

$$\bar{a}_i^T x - \left\| P_i^T x \right\|_2 \geq b_i \ , \qquad i = 1,\ldots,N_c$$

$$\bar{\lambda}^T x + \left\| P_0^T x \right\|_2 \leq d \ ,$$

the rest of the objectives and constraints are the same as in the deterministic problem.

6.3.1 Realizations of the hydraulic conductivity

The uncertainty of the hydraulic conductivity is taken into consideration. For generating the random values of the hydraulic conductivity Sequential Gaussian Simulation (SGSIM) of the GSLIB is proposed to simulate 20 realizations of hydraulic conductivity values. The mean and variance of the random field and its two-dimensional result of these realizations is graphically shown in the Figures 6.13, 6.14 and 6.15.

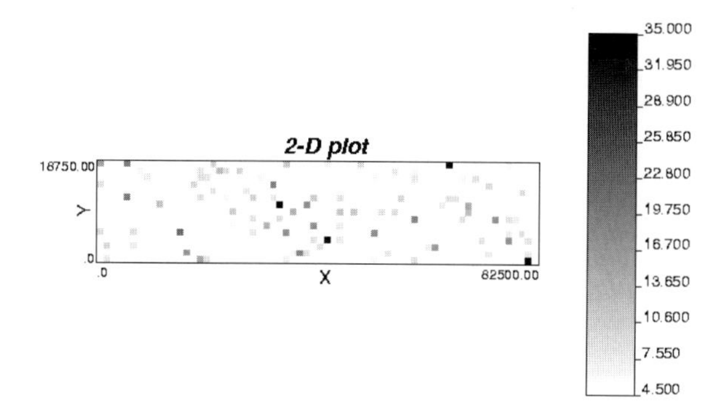

Figure 6.13: The 2-D distribution of the first realization of hydraulic conductivity.

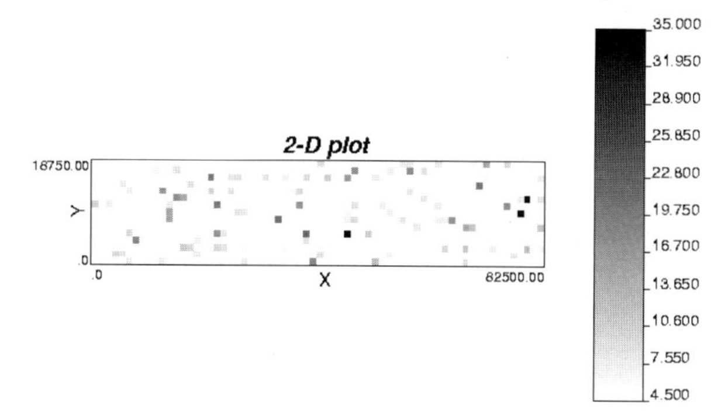

Figure 6.14: The 2-D distribution of the twentieth realization of hydraulic conductivity.

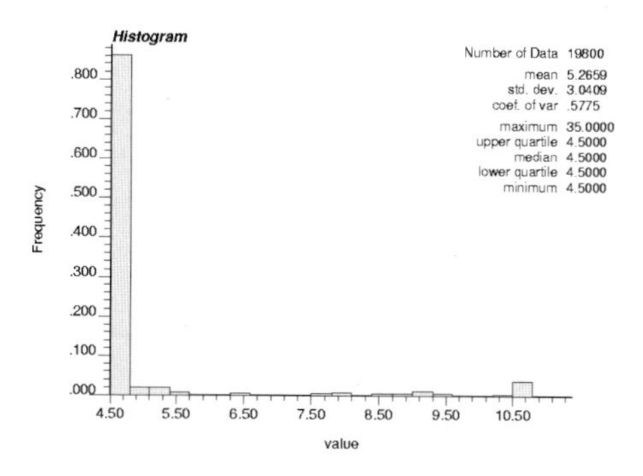

Figure 6.15: The histogram of the random data for 20 realizations of hydraulic conductivity.

6.3.2 The results from the method of minimum distance from the ideal solution

In chapter 5, the form of the robust multi-objective optimization problem for the SW intrusion management formulated by the L_2-metric method was described as follows:

$$\text{Max} \quad -t_2 \tag{5.71}$$

s.t.

$$\frac{t+1}{2} \geq \left\| \left(\frac{t-1}{2}, \beta, c^T x - \overline{Z}_1 \right)^T \right\|_2, \tag{5.73'}$$

$$\overline{Z}_2 + \beta - \overline{\lambda}^T x \geq \left\| \eta P_0^T x \right\|_2, \tag{5.74'}$$

$$-\ b_i + \overline{a}_i^T x \geq \left\| \eta P_i^T x \right\|_2, \qquad i = 1,\ldots,N_c \tag{5.51'}$$

$$d - \overline{\lambda}^T x \geq \left\| \eta P_0^T x \right\|_2, \tag{5.52'}$$

$$-W_D + e^T x \geq 0, \tag{5.5}$$

$$Wc_i - x_i \geq 0, \qquad i = 1,\ldots,N_{iw}+N_{ew} \tag{5.6}$$

$$x_i \geq 0, \qquad i = 1,\ldots,N_{iw}+N_{ew} \tag{5.7}$$

Results

1. The optimal solution $x = \boldsymbol{Q}_{opt}$

The optimal solutions of the L_2-metric problem of all iterations which are shown in Table 6.7 are based on the response matrices. These are created by the perturbed pumping rate ΔQ which is initialized from 10 m^3/h and gradually decreased by steps of 0.2 m^3/h. In this problem, the twenty average values of the simulated saltwater intrusion increments are computed and reported for each iteration.

2. The optimal values of Z_1 and Z_2 and the mean values of SW intrusion increments, *Lsim-avg*, for 20 realizations of the hydraulic conductivity, which are shown in Table 6.8, are based on the optimal solution at different iterations.

For this L_2-metric problem, the optimal value of Z_2 (–1698.58 m) is much greater than the one (–2669.87 m) in the deterministic case whereas the optimal value of Z_1 (643 MU/h) is just slightly smaller than the one (648 MU/h) in the deterministic case.

The non-inferior sets drawn in Figures 6.16 and 6.17 are created by the constraint method and based on the optimal solution from the L_2-metric problem. The optimal values with $\alpha = 1$ and $\alpha = \infty$ are just for showing the narrow range of choice in the non-inferior set for the L_α-metric problem. In Figure 6.17 the trade-off of the optimal values in the narrow range of the non-inferior set can be seen, when α is changed from 1 to infinity in the L_α-metric problems.

The trade-off values of (Z_1, Z_2) in the non-inferior set at the first and last iterations are depicted in Figures 6.16 and 6.17.

3. The optimal solution $x = \boldsymbol{Q}_{opt}$, which is shown in Table 6.9, is obtained at the ninth iteration (k = 9). Unlike the deterministic problems, this L_2-metric optimal solution is not symmetric when the injection rate (851.5 m^3/h) of cell (28, 2) is slightly different from the injection rate (900 m^3/h) of the same cell in the deterministic case. This is caused by the heterogeneity of the random values of the distributed transmissivities of the aquifer. These values of $\boldsymbol{Q}_{opt}$ are used for the input

file of SHARP and Figure 6.20 demonstrates one of 20 output files of SHARP for mapping the extent of SW intrusion, based on the last-iteration optimal solution.

4. The SW intrusion interface simulated by SHARP computer code

For the L_2-metric problem, the simulated maps of saltwater toe and freshwater tip above are not symmetric. The reasons for the asymmetry of the interfaces are, mainly, the asymmetric optimal solution and the heterogeneity of the aquifer transmissivities by the uncertainty.

The mean value, −1857.18 m, which is computed from twenty simulated saltwater intrusion increments varying in the range of (−1795.18, −1895.93), (see Table 6.7), is about 9% smaller than the optimal value Z_2 (−1698.58 m).

Discussion

For the L_2-metric problem, the random values of the distributed transmissivities of aquifer cause the asymmetry of the optimal solution. On the other hand, also because of this uncertainty concerning aquifer transmissivities the saltwater intrusion increment is much greater than the one in the deterministic case. At the same time the operational costs are more or less the same for the two cases. This means that with the same optimal allocation of wells, their operation under the uncertainty case will take more risks than in the deterministic case in the sense of the salt/fresh interfaces which are closer to the capture zone of these wells.

In Figure 6.17 we can see the distinct difference between two parts of the non-inferior set. The point (643.14, −1698.57) is the intersection point of the two parts. Below this point the MRT will be smaller.

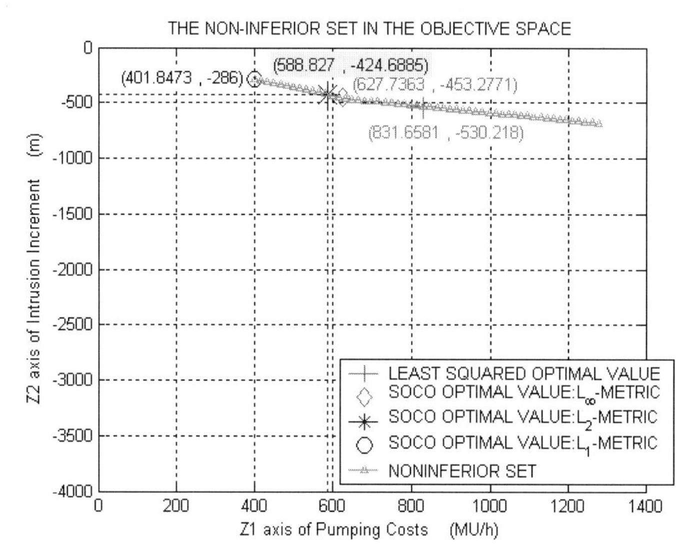

Figure 6.16: The non-inferior set in the objective space at the first iteration

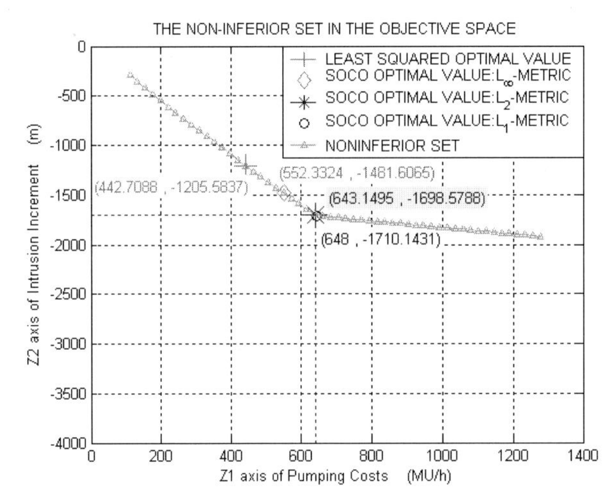

Figure 6.17: The non-inferior set in the objective space at the ninth iteration.

```
MAP OF EXTENT OF INTRUSION (F-FRESHWATER. M-FRESH AND SALTWATER. S-SALTWATER)
                 10        20        30        40        50        60          146
        12345678901234567890123456789012345678901345678901234567890123456789012345  12456
   15   .................................................................  .....
   14   .FRRRRRRRRREFFFFEFFFFFIFFFFIMMMMMMMMSSSSSSSSSSSSSSSSSSSSSSSSSSSSS  SSSSS
   13   .FRRRRRRRRRFFFFFFFFFFFFFFFFFMMMMMMMMSSSSSSSSSSSSSSSSSSSSSSSSSSSSS  SSSSS
   12   .FRRRRRRRRREFFFFEFFFFFIFFFFIMMMMMMMMSSSSSSSSSSSSSSSSSSSSSSSSSSSSS  SSSSS
   11   .FRRRRRRRRRFFFFFFFFFFFFFFFFFMMMMMMMMSSSSSSSSSSSSSSSSSSSSSSSSSSSSS  SSSSS
   10   .FRRRRRRRRREFFFFEFFFFFIFFFFIMMMMMMMMSSSSSSSSSSSSSSSSSSSSSSSSSSSSS  SSSSS
    9   .FRRRRRRRRRFFFFFFFFFFFFFFFFFMMMMMMMMSSSSSSSSSSSSSSSSSSSSSSSSSSSSS  SSSSS
    8   .FRRRRRRRRREFFFFEFFFFFIFFFFIMMMMMMMMSSSSSSSSSSSSSSSSSSSSSSSSSSSSS  SSSSS
    7   .FRRRRRRRRRFFFFFFFFFFFFFFFFFMMMMMMMMSSSSSSSSSSSSSSSSSSSSSSSSSSSSS  SSSSSS
    6   .FRRRRRRRRREFFFFEFFFFFIFFFFIMMMMMMMMSSSSSSSSSSSSSSSSSSSSSSSSSSSSS  SSSSSS
    5   .FRRRRRRRRRFFFFFFFFFFFFFFFFFMMMMMMMMSSSSSSSSSSSSSSSSSSSSSSSSSSSSS  SSSSSS
    4   .FRRRRRRRRREFFFFEFFFFFIFFFFIMMMMMMMMSSSSSSSSSSSSSSSSSSSSSSSSSSSSS  SSSSSS
    3   .FRRRRRRRRRFFFFFFFFFFFFFFFFFMMMMMMMMSSSSSSSSSSSSSSSSSSSSSSSSSSSSS  SSSSSS
    2   .FRRRRRRRRREFFFFEFFFFFIFFFFIMMMMMMMMSSSSSSSSSSSSSSSSSSSSSSSSSSSSS  SSSSSS
    1   .................................................................  .....
        12345678901234567890123456789012345678901345678901234567890123456789012345  23456
                 10        20        30        40        50        60          146
(NOTE: R-RESIDENTIAL AREA. E- CANDIDATE EXTRACTION. I- CANDIDATE INJECTION)
```

Figure 6.18: The SW toe and FW tip at the non-pumping condition.

```
MAP OF EXTENT OF INTRUSION (F-FRESHWATER. M-FRESH AND SALTWATER. S-SALTWATER)
                 10        20        30        40        50        60          146
        12345678901234567890123456789012345678901345678901234567890123456789012345  12456
   15   .................................................................  .....
   14   .FRRRRRRRRREFFFFEFFFFFIFFFFIFFFFFFFFMMMMMMMMSSSSSSSSSSSSSSSSSSSSS  SSSSS
   13   .FRRRRRRRRRFFFFFFFFFFFFFFFFFFFFFFFMMMMMMMMMSSSSSSSSSSSSSSSSSSSSSS  SSSSS
   12   .FRRRRRRRRREFFFFEFFFFFIFFFFIFFFFFFFMMMMMMMMMSSSSSSSSSSSSSSSSSSSSS  SSSS
   11   .FRRRRRRRRRFFFFFFFFFFFFFFFFFFFFFFMMMMMMMMMMSSSSSSSSSSSSSSSSSSSSSS  SSSSS
   10   .FRRRRRRRRREFFFFEFFFFFIFFFFIFFFFFMMMMMMMMMSSSSSSSSSSSSSSSSSSSSSSS  SSSSS
    9   .FRRRRRRRRRFFFFFFFFFFFFFFFFFFFFFFFMMMMMMMMMSSSSSSSSSSSSSSSSSSSSSS  SSSSS
    8   .FRRRRRRRRREFFFFEFFFFFIFFFFIFFFFFFFMMMMMMMMMSSSSSSSSSSSSSSSSSSSSS  SSSSS
    7   .FRRRRRRRRRFFFFFFFFFFFFFFFFFFFFFFFMMMMMMMMMSSSSSSSSSSSSSSSSSSSSSS  SSSSSS
    6   .FRRRRRRRRREFFFFEFFFFFIFFFFIFFFFFFFMMMMMMMMMSSSSSSSSSSSSSSSSSSSSS  SSSSSS
    5   .FRRRRRRRRRFFFFFFFFFFFFFFFFFFFFFFFFMMMMMMMMMSSSSSSSSSSSSSSSSSSSSS  SSSSSS
    4   .FRRRRRRRRREFFFFEFFFFFIFFFFIFFFFFFFMMMMMMMMMMSSSSSSSSSSSSSSSSSSSS  SSSSSS
    3   .FRRRRRRRRRFFFFFFFFFFFFFFFFFFFFFFFFMMMMMMMMMMSSSSSSSSSSSSSSSSSSSS  SSSSSS
    2   .FRRRRRRRRREFFFFEFFFFFIFFFFIFFFFFFFFFMMMMMMMMSSSSSSSSSSSSSSSSSSSS  SSSSSS
    1   .................................................................  .....
        12345678901234567890123456789012345678901345678901234567890123456789012345  23456
                 10        20        30        40        50        60          146
NOTE: R-RESIDENTIAL AREA, E-CANDIDATE EXTRACTION, I- CANDIDATE INJECTION, E-OPTIMAL EXTRACTION, I-OPTIMAL INJECTION
```

Figure 6.19: The SW toe and FW tip after the first iteration.

```
MAP OF EXTENT OF INTRUSION (F-FRESHWATER. M-FRESH AND SALTWATER. S-SALTWATER)
                   10        20        30        40       50        60        146
         12345678901234567890123456789012345678901345678901234567890123456789012345  12456
    15   .................................................................  .....
    14   .FRRRRRRRRRREFFFFEFFFFFIFFFFIFFFFFFFFMMMMMMMMMSSSSSSSSSSSSSSSSSSSS  SSSSS
    13   .FRRRRRRRRRFFFFFFFFFFFFFFFFFFFFFFFFMMMMMMMMMMSSSSSSSSSSSSSSSSSSSS  SSSSS
    12   .FRRRRRRRRRREFFFFEFFFFFIFFFFIFFFFFFFMMMMMMMMMMSSSSSSSSSSSSSSSSSSSS  SSSSS
    11   .FRRRRRRRRRFFFFFFFFFFFFFFFFFFFFFFFFMMMMMMMMMMSSSSSSSSSSSSSSSSSSSS  SSSSS
    10   .FRRRRRRRRRREFFFFEFFFFFIFFFFIFFFFFFFMMMMMMMMMMSSSSSSSSSSSSSSSSSSSS  SSSSS
     9   .FRRRRRRRRRFFFFFFFFFFFFFFFFFFFFFFFFMMMMMMMMMMSSSSSSSSSSSSSSSSSSSS  SSSSS
     8   .FRRRRRRRRRREFFFFEFFFFFIFFFFIFFFFFFFFMMMMMMMMMSSSSSSSSSSSSSSSSSSSS  SSSSS
     7   .FRRRRRRRRRFFFFFFFFFFFFFFFFFFFFFFFFMMMMMMMMMMSSSSSSSSSSSSSSSSSSSS  SSSSS
     6   .FRRRRRRRRRREFFFFEFFFFFIFFFFIFFFFFFFMMMMMMMMMMSSSSSSSSSSSSSSSSSSSS  SSSSS
     5   .FRRRRRRRRRFFFFFFFFFFFFFFFFFFFFFFFFMMMMMMMMMMSSSSSSSSSSSSSSSSSSSS  SSSSS
     4   .FRRRRRRRRRREFFFFEFFFFFIFFFFIFFFFFFFMMMMMMMMMMSSSSSSSSSSSSSSSSSSSS  SSSSS
     3   .FRRRRRRRRRFFFFFFFFFFFFFFFFFFFFFFFFMMMMMMMMMMSSSSSSSSSSSSSSSSSSSS  SSSSS
     2   .FRRRRRRRRRREFFFFEFFFFFIFFFFIFFFFFFFFMMMMMMMMMSSSSSSSSSSSSSSSSSSSS  SSSSS
     1   .................................................................  .....
         12345678901234567890123456789012345678901345678901234567890123456789012345  23456
                   10        20        30        40       50        60        146
NOTE: R-RESIDENTIAL AREA, E-CANDIDATE EXTRACTION, I- CANDIDATE INJECTION, E-OPTIMAL EXTRACTION, I-OPTIMAL INJECTION
```

Figure 6.20: The SW toe and FW tip after the ninth iteration.

Table 6.7: The results of the L_2-metric problem under uncertainty case.

Iteration, k	1	2	3	4	5	6	7	8	9
ΔQ (m^3/h)	10.00	9.80	9.60	9.40	9.20	9.00	8.80	8.60	8.400
Z_1 (MU/h)	588.83	641.83	644.35	643.30	641.60	644.46	643.09	643.16	643.15
Z_2 (m)	-424.69	-1717.72	-1718.50	-1702.83	-1695.51	-1718.84	-1699.35	-1707.30	-1698.58
$Lsim_1$ (m)	-1751.54	-1887.14	-1890.99	-1889.45	-1886.90	-1891.08	-1888.52	-1889.26	-1889.23
:	-1744.53	-1862.37	-1866.15	-1864.68	-1862.02	-1866.33	-1864.28	-1864.45	-1864.37
:	-1741.46	-1866.00	-1871.62	-1869.12	-1865.88	-1871.63	-1868.75	-1868.54	-1868.51
:	-1727.79	-1855.01	-1859.97	-1858.01	-1854.55	-1860.19	-1857.58	-1857.74	-1857.71
:	-1756.01	-1861.03	-1867.77	-1865.25	-1860.58	-1868.04	-1864.56	-1864.77	-1864.73
:	-1728.15	-1870.55	-1874.39	-1872.42	-1870.13	-1874.32	-1872.22	-1872.58	-1872.37
:	-1708.61	-1839.85	-1844.01	-1842.39	-1839.35	-1843.78	-1842.16	-1842.42	-1842.23
:	-1729.21	-1858.22	-1861.75	-1860.19	-1856.46	-1861.95	-1859.86	-1859.98	-1859.96
:	-1743.86	-1858.08	-1861.36	-1860.29	-1857.76	-1861.83	-1859.65	-1859.62	-1859.60
:	-1748.97	-1867.04	-1870.66	-1869.03	-1866.52	-1870.82	-1868.71	-1868.82	-1868.80
:	-1789.26	-1893.68	-1897.82	-1896.17	-1893.34	-1898.00	-1895.83	-1895.95	-1895.93
:	-1746.80	-1871.87	-1875.61	-1874.15	-1871.52	-1875.90	-1873.82	-1873.96	-1873.92
:	-1722.03	-1840.80	-1845.07	-1843.10	-1840.76	-1845.26	-1842.68	-1842.94	-1842.87
:	-1719.69	-1841.18	-1845.39	-1843.15	-1840.91	-1845.72	-1843.22	-1843.35	-1843.33
:	-1710.88	-1838.91	-1843.43	-1842.08	-1838.59	-1843.74	-1841.44	-1841.36	-1841.12
:	-1730.53	-1854.04	-1857.64	-1856.17	-1853.66	-1858.10	-1856.04	-1856.05	-1856.20
:	-1711.40	-1826.09	-1830.61	-1828.82	-1825.84	-1830.71	-1828.40	-1828.32	-1828.49
:	-1751.81	-1873.46	-1879.54	-1877.18	-1873.40	-1879.79	-1876.66	-1876.84	-1876.81
:	-1721.99	-1840.93	-1844.67	-1842.25	-1840.46	-1844.30	-1842.08	-1842.14	-1842.22
$Lsim_{20}$ (m)	-1675.74	-1793.20	-1797.10	-1795.16	-1792.92	-1797.37	-1795.03	-1795.20	-1795.18
$\|(Q_i^{k} - Q_i^{k-1})\|$	10000.00	594.94	25.16	10.47	17.02	28.63	13.74	0.77	0.16
Q_1 (m^3/h)	50.00	50.00	50.00	50.00	50.00	50.00	50.00	50.00	50.00
:	50.00	50.00	50.00	50.00	50.00	50.00	50.00	50.00	50.00
:	50.00	50.00	50.00	50.00	50.00	50.00	50.00	50.00	50.00
:	50.00	50.00	50.00	50.00	50.00	50.00	50.00	50.00	50.00
:	50.00	50.00	50.00	50.00	50.00	50.00	50.00	50.00	50.00
:	50.00	50.00	50.00	50.00	50.00	50.00	50.00	50.00	50.00
:	50.00	50.00	50.00	50.00	50.00	50.00	50.00	50.00	50.00
:	0.00	0.00	0.00	0.00	0.00	0.00	0.00	0.00	0.00
:	0.00	0.00	0.00	0.00	0.00	0.00	0.00	0.00	0.00
:	0.00	0.00	0.00	0.00	0.00	0.00	0.00	0.00	0.00
:	10.00	10.00	10.00	10.00	10.00	10.00	10.00	10.00	10.00

:	0.00	0.00	0.00	0.00	0.00	0.00	0.00	0.00	0.00
:	0.00	0.00	0.00	0.00	0.00	0.00	0.00	0.00	0.00
:	0.00	0.00	0.00	0.00	0.00	0.00	0.00	0.00	0.00
:	0.00	0.00	0.00	0.00	0.00	0.00	0.00	0.00	0.00
:	0.00	0.00	0.00	0.00	0.00	0.00	0.00	0.00	0.00
:	0.00	0.00	0.00	0.00	0.00	0.00	0.00	0.00	0.00
:	0.00	0.00	0.00	0.00	0.00	0.00	0.00	0.00	0.00
:	0.00	0.00	0.00	0.00	0.00	0.00	0.00	0.00	0.00
:	0.00	0.00	0.00	0.00	0.00	0.00	0.00	0.00	0.00
:	0.00	0.00	0.00	0.00	0.00	0.00	0.00	0.00	0.00
:	900.00	838.33	863.48	853.01	835.99	864.62	850.88	851.65	851.49
:	900.00	900.00	900.00	900.00	900.00	900.00	900.00	900.00	900.00
:	900.00	900.00	900.00	900.00	900.00	900.00	900.00	900.00	900.00
:	900.00	900.00	900.00	900.00	900.00	900.00	900.00	900.00	900.00
:	308.27	900.00	900.00	900.00	900.00	900.00	900.00	900.00	900.00
:	900.00	900.00	900.00	900.00	900.00	900.00	900.00	900.00	900.00
Q_{28} (m^3/h)	900.00	900.00	900.00	900.00	900.00	900.00	900.00	900.00	900.00

Table 6.8: The optimal values and the *Lsim-avg* at all different iterations (L_2-metric).

Iterations	ΔQ (m^3/h)	Z_1 (MU/h)	Z_2 (m)	*Lsim-avg* (m)	$\|(Q_i^{\,k} - Q_i^{\,k-1})\|$
1	10.0	588.83	-424.69	-1733.012	10000.000
2	9.8	641.83	-1717.72	-1854.972	594.935
3	9.6	644.35	-1718.50	-1859.277	25.155
4	9.4	643.30	-1702.83	-1857.453	10.472
5	9.2	641.60	-1695.51	-1854.577	17.019
6	9.0	644.46	-1718.84	-1859.443	28.625
7	8.8	643.09	-1699.35	-1857.074	13.738
8	8.6	643.16	-1707.30	-1857.215	0.770
9	8.4	643.15	-1698.58	-1857.178	0.155

Table 6.9: The optimal solution at the last iteration (L_2-metric)

$\boldsymbol{Q}_{opt}$ (m^3/h)	Column 12	Column 17	Column 23	Column 28
Row 14	50.00	0.00	0.00	900.00
Row 12	50.00	0.00	0.00	900.00
Row 10	50.00	0.00	0.00	900.00
Row 8	50.00	10.00	0.00	900.00
Row 6	50.00	0.00	0.00	900.00
Row 4	50.00	0.00	0.00	900.00
Row 2	50.00	0.00	0.00	851.49

6.3.3 The results from the method of prior assessments of weights (weighted problem)

Here another optimal solution is shown by using the method of prior assessments of weights for the robust multi-objective problem of SW intrusion management. With w = 0.1, 0.6, 3.1 > 0, different optimal solutions can be obtained for the multi-objective problem from (5.63), (5.75) and (5.61) written in the form (see Chapter 5):

$$\text{Min} \quad t_w \tag{5.35}$$

s.t.

$$t_w \geq Z_1 + wZ_2 = c^T x + w\left(\bar{\lambda}^T x + \left\|\eta P_0^T x\right\|_2\right) \tag{5.75}$$

$$x \in \boldsymbol{\Omega} \tag{5.61}$$

Results

1. The optimal solution $x = \boldsymbol{Q}_{opt}$

The optimal solutions and the optimal values of three iterations are shown in Table 6.10. ΔQ is initially equal to 10 m^3/h and gradually decreased by steps of 0.2 m^3/h. In the uncertainty case, the $w = 0.6$ weighted problem converges to the optimal solution at the third iteration.

2. The optimal (trade-off) values of Z_1 and Z_2 and the mean value of SW intrusion increments, *Lsim_avg*, for 20 realizations of the hydraulic conductivity are shown in Table 6.11. The trade-off values of (Z_1, Z_2) in the non-inferior set at the first and last iteration are depicted in Figures 6.21 and 6.22.

The non-inferior sets drawn in Figure 6.21 and 6.22 are created by the constraint method. These are based on the optimal solution from the $w = 0.6$ weighted problem. The optimal values with the weights $w = 0.1$ and $w = 3.1$ are just for showing the full range of choice in the non-inferior set for the weighted problem. In Figure 6.22 the trade-off can be seen of the optimal values along the whole non-inferior set when altering the weight values, from, for example, 0.1 to 3.1.

3. The optimal solution $x = \boldsymbol{Q}_{opt}$, which is shown in Table 6.12, is obtained at the third iteration ($k = 3$). The reader can refer these values of $\boldsymbol{Q}_{opt}$ to Figure 6.25 that is the output file of SHARP for mapping the extent of SW intrusion based on the last-iteration optimal solution vector.

4. The SW intrusion interface simulated by SHARP computer code

The maps in Figures 6.23, 6.24 and 6.25 show the asymmetry of the saltwater toe and freshwater tip. At the last iteration, the mean value of the twenty simulated saltwater intrusion increments is computed as −1317.71 m. The simulated values, which are reported in Table 6.10, vary in the range of −1276.65 to −1414.70. The mean value (−1317.71 m) is 7.73% smaller than the optimal value (−1223.21), which shows the small difference between the optimal value and simulated value.

Discussion

For the weighted problem ($w = 0.6$), its optimal solution under the uncertainty case is not symmetric if compared to the deterministic case. Hence, the saltwater intrusion increments at all rows are not symmetric either. This can be explained by the heterogeneity of the aquifer under the uncertainty.

For the uncertainty case, the trade-off can also be seen among the optimal values (Z_1, Z_2) between two methods, i.e. a trade-off between the optimal values (643, −1698.58) and (468, −1223.21) of the L_2-metric and weighted problems, respectively.

The uncertainty of the aquifer transmissivities also makes the interfaces closer to the capture zone of the same optimal wells if compared to the deterministic case. This also implies that the risk due to the uncertainty of the aquifer parameter is taken into the computation. Hence, the average saltwater intrusion increment is predicted to be greater than in the deterministic case where there is no risk of the parameter uncertainty.

In Figure 6.22 we can see the difference between two parts of the non-inferior set. The point (468, −1223.20) is the intersection point of the two parts. Below this point the MRT will gradually decrease.

Table 6.10: The results of the weighted problem, $w = 0.6$, under uncertainty case.

Iteration, k	1	2	3
ΔQ (m^3/h)	10.00	9.80	9.60
Z_1 (MU/h)	401.85	468.00	468.00
Z_2 (m)	-286.00	-1162.16	-1223.21
$Lsim_1$ (m)	-1187.95	-1330.87	-1330.87
:	-1171.17	-1318.33	-1318.33
:	-1163.88	-1319.36	-1319.36
:	-1170.12	-1309.55	-1309.55
:	-1143.86	-1302.88	-1302.88
:	-1154.67	-1310.60	-1310.60
:	-1137.02	-1288.51	-1288.51
:	-1175.15	-1322.05	-1322.05
:	-1180.94	-1326.81	-1326.81
:	-1169.49	-1329.61	-1329.61
:	-1247.01	-1414.70	-1414.70
:	-1161.89	-1323.58	-1323.58
:	-1152.09	-1299.99	-1299.99
:	-1161.36	-1304.88	-1304.88
:	-1143.87	-1295.62	-1295.62
:	-1143.59	-1324.89	-1324.89
:	-1181.92	-1322.24	-1322.24
:	-1165.86	-1336.01	-1336.01
:	-1152.96	-1297.10	-1297.10
$Lsim_{20}$ (m)	-1137.99	-1276.65	-1276.65
$\|(Q_i^k - Q_i^{k-1})\|$	10000.00	661.53	0.00
Q_1 (m^3/h)	50.00	50.00	50.00
:	50.00	50.00	50.00
:	50.00	50.00	50.00
:	50.00	50.00	50.00
:	50.00	50.00	50.00
:	50.00	50.00	50.00
:	50.00	50.00	50.00
:	0.00	0.00	0.00
:	0.00	0.00	0.00
:	0.00	0.00	0.00
:	10.00	10.00	10.00
:	0.00	0.00	0.00
:	0.00	0.00	0.00
:	0.00	0.00	0.00
:	0.00	0.00	0.00
:	0.00	0.00	0.00
:	0.00	0.00	0.00
:	0.00	0.00	0.00
:	0.00	0.00	0.00
:	0.00	0.00	0.00
:	0.00	0.00	0.00
:	900.00	900.00	900.00
:	238.47	900.00	900.00
:	900.00	900.00	900.00
:	0.00	0.00	0.00
:	0.00	0.00	0.00
:	900.00	900.00	900.00
Q_{28} (m^3/h)	900.00	900.00	900.00

Table 6.11: The optimal values and the *Lsim-avg* at all different iterations ($w = 0.6$).

Iteration	ΔQ (m^3/h)	Z_1 (MU/h)	Z_2 (m)	*Lsim_avg* (m)	$\|(Q_i^k - Q_i^{k-1})\|$
1	10	401.85	-286.00	-1165.13	10000.00
2	9.8	468.00	-1162.16	-1317.71	661.53
3	9.6	468.00	-1223.21	-1317.71	1.58E-07

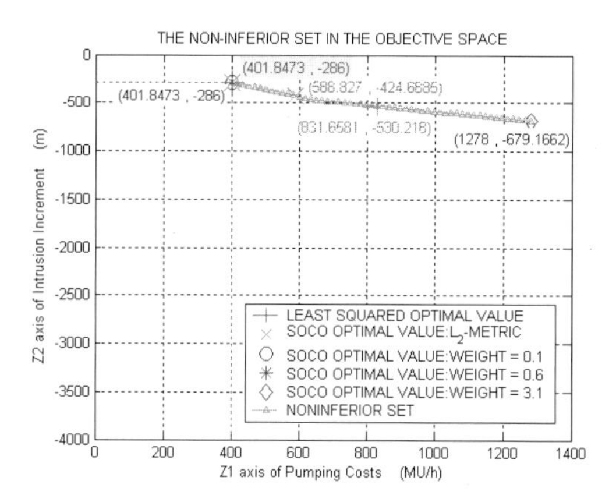

Figure 6.21: The non-inferior set in the objective space at the first iteration.

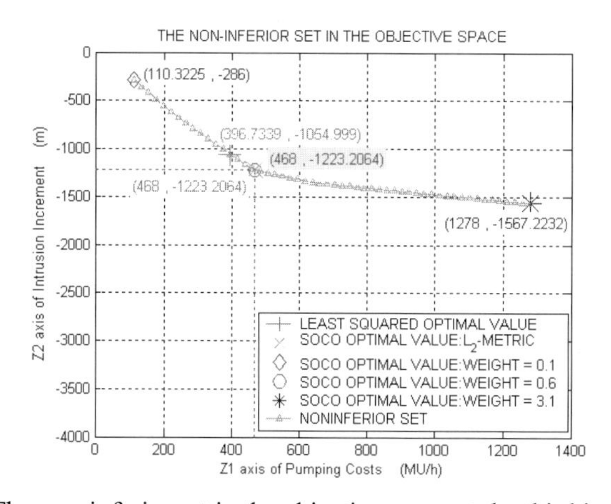

Figure 6.22: The non-inferior set in the objective space at the third iteration.

Table 6.12: The optimal solution at the last iteration ($w = 0.6$)

Q_{opt} (m^3/h)	Column 12	Column 17	Column 23	Column 28
Row 14	50.00	0.00	0.00	900.00
Row 12	50.00	0.00	0.00	900.00
Row 10	50.00	0.00	0.00	0.00
Row 8	50.00	10.00	0.00	0.00
Row 6	50.00	0.00	0.00	900.00
Row 4	50.00	0.00	0.00	900.00
Row 2	50.00	0.00	0.00	900.00

```
MAP OF EXTENT OF INTRUSION (F-FRESHWATER. M-FRESH AND SALTWATER. S-SALTWATER)
                 10        20        30        40        50        60          146
         12345678901234567890123456789012345678901345678901234567890123451  12456
   15    .................................................................  .....
   14    .FRRRRRRRRREFFFFEFFFFFIFFFFIMMMMMMMMMSSSSSSSSSSSSSSSSSSSSSSSSSSSS  SSSSS
   13    .FRRRRRRRRRFFFFFFFFFFFFFFFFFMMMMMMMMMSSSSSSSSSSSSSSSSSSSSSSSSSSSS  SSSSS
   12    .FRRRRRRRRREFFFFEFFFFFIFFFFIMMMMMMMMMSSSSSSSSSSSSSSSSSSSSSSSSSSSS  SSSSS
   11    .FRRRRRRRRRFFFFFFFFFFFFFFFFFMMMMMMMMMSSSSSSSSSSSSSSSSSSSSSSSSSSSS  SSSSS
   10    .FRRRRRRRRREFFFFEFFFFFIFFFFIMMMMMMMMMSSSSSSSSSSSSSSSSSSSSSSSSSSSS  SSSSS
    9    .FRRRRRRRRRFFFFFFFFFFFFFFFFFMMMMMMMMMSSSSSSSSSSSSSSSSSSSSSSSSSSSS  SSSSS
    8    .FRRRRRRRRREFFFFEFFFFFIFFFFIMMMMMMMMMSSSSSSSSSSSSSSSSSSSSSSSSSSS   SSSSS
    7    .FRRRRRRRRRFFFFFFFFFFFFFFFFFMMMMMMMMMSSSSSSSSSSSSSSSSSSSSSSSSSSS   SSSSSS
    6    .FRRRRRRRRREFFFFEFFFFFIFFFFIMMMMMMMMMSSSSSSSSSSSSSSSSSSSSSSSSSSS   SSSSSS
    5    .FRRRRRRRRRFFFFFFFFFFFFFFFFFMMMMMMMMMSSSSSSSSSSSSSSSSSSSSSSSSSSS   SSSSSS
    4    .FRRRRRRRRREFFFFEFFFFFIFFFFIMMMMMMMMMSSSSSSSSSSSSSSSSSSSSSSSSSSS   SSSSSS
    3    .FRRRRRRRRRFFFFFFFFFFFFFFFFFMMMMMMMMMSSSSSSSSSSSSSSSSSSSSSSSSSSS   SSSSSS
    2    .FRRRRRRRRREFFFFEFFFFFIFFFFIMMMMMMMMMSSSSSSSSSSSSSSSSSSSSSSSSSSS   SSSSSS
    1    .................................................................  .....
         12345678901234567890123456789012345678901345678901234567890123451  23456
                 10        20        30        40        50        60          146
NOTE: R-RESIDENTIAL AREA. E- CANDIDATE EXTRACTION. I- CANDIDATE INJECTION
```

Figure 6.23: The SW toe and FW tip at the non-pumping condition simulated with the twentieth realization of the hydraulic.

```
   MAP OF EXTENT OF INTRUSION (F-FRESHWATER. M-FRESH AND SALTWATER. S-SALTWATER)
                 10        20        30        40        50        60          146
         12345678901234567890123456789012345678901345678901234567890123451  12456
   15    .................................................................  .....
   14    .FRRRRRRRRREFFFFEFFFFFIFFFFIFFFFFFFFFMMMMMMMMMSSSSSSSSSSSSSSSSSSSS  SSSSS
   13    .FRRRRRRRRRFFFFFFFFFFFFFFFFFFFFFFFFFFMMMMMMMMMSSSSSSSSSSSSSSSSSSSS  SSSSS
   12    .FRRRRRRRRREFFFFEFFFFFIFFFFIFFFFFFFFMMMMMMMMMSSSSSSSSSSSSSSSSSSSSS   SSSS
   11    .FRRRRRRRRRFFFFFFFFFFFFFFFFFFFFFFFFMMMMMMMMMMSSSSSSSSSSSSSSSSSSSSS  SSSSS
   10    .FRRRRRRRRREFFFFEFFFFFIFFFFIFMMMMMMMMMMMSSSSSSSSSSSSSSSSSSSSSSSSS  SSSSS
    9    .FRRRRRRRRRFFFFFFFFFFFFFFFFFFMMMMMMMMMMSSSSSSSSSSSSSSSSSSSSSSSSSSS  SSSSS
    8    .FRRRRRRRRREFFFFEFFFFFIFFFFIFMMMMMMMMMMMMSSSSSSSSSSSSSSSSSSSSSSSS   SSSSSS
    7    .FRRRRRRRRRFFFFFFFFFFFFFFFFFFFFFFFFMMMMMMMMMMSSSSSSSSSSSSSSSSSSSS   SSSSSS
    6    .FRRRRRRRRREFFFFEFFFFFIFFFFIFFFFFFMMMMMMMMMMSSSSSSSSSSSSSSSSSSSSS   SSSSSS
    5    .FRRRRRRRRRFFFFFFFFFFFFFFFFFFFFFFFFFMMMMMMMMMSSSSSSSSSSSSSSSSSSSS   SSSSSS
    4    .FRRRRRRRRREFFFFEFFFFFIFFFFIFFFFFMMMMMMMMMMSSSSSSSSSSSSSSSSSSSSSS   SSSSSS
    3    .FRRRRRRRRRFFFFFFFFFFFFFFFFFFFFFFFFFFMMMMMMMMMSSSSSSSSSSSSSSSSSSS   SSSSSS
    2    .FRRRRRRRRREFFFFEFFFFFIFFFFIFFFFFFFFFMMMMMMMMMSSSSSSSSSSSSSSSSSSS   SSSSSS
    1    .................................................................  .....
         12345678901234567890123456789012345678901345678901234567890123451  23456
                 10        20        30        40        50        60          146
NOTE: R-RESIDENTIAL AREA, E-CANDIDATE EXTRACTION, I- CANDIDATE INJECTION, E-OPTIMAL EXTRACTION, I-OPTIMAL INJECTION
```

Figure 6.24: The SW toe and FW tip simulated with the twentieth realization of the hydraulic conductivity after the first iteration.

```
   MAP OF EXTENT OF INTRUSION (F-FRESHWATER. M-FRESH AND SALTWATER. S-SALTWATER)
                 10        20        30        40        50        60          146
         12345678901234567890123456789012345678901345678901234567890123451  12456
   15    .................................................................  .....
   14    .FRRRRRRRRREFFFFEFFFFFIFFFFIFFFFFFFFFMMMMMMMMMSSSSSSSSSSSSSSSSSSSS  SSSSS
   13    .FRRRRRRRRRFFFFFFFFFFFFFFFFFFFFFFFFFFMMMMMMMMMSSSSSSSSSSSSSSSSSSSS  SSSSS
   12    .FRRRRRRRRREFFFFEFFFFFIFFFFIFFFFFFFMMMMMMMMMMSSSSSSSSSSSSSSSSSSSSS   SSSS
   11    .FRRRRRRRRRFFFFFFFFFFFFFFFFFFFFFFFFMMMMMMMMMMSSSSSSSSSSSSSSSSSSSSS  SSSSS
   10    .FRRRRRRRRREFFFFEFFFFFIFFFFIFMMMMMMMMMMMSSSSSSSSSSSSSSSSSSSSSSSSS  SSSSS
    9    .FRRRRRRRRRFFFFFFFFFFFFFFFFFFMMMMMMMMMMSSSSSSSSSSSSSSSSSSSSSSSSSSS  SSSSS
    8    .FRRRRRRRRREFFFFEFFFFFIFFFFIFMMMMMMMMMMMMSSSSSSSSSSSSSSSSSSSSSSSS   SSSSSS
    7    .FRRRRRRRRRFFFFFFFFFFFFFFFFFFFFFFFFMMMMMMMMMMSSSSSSSSSSSSSSSSSSSS   SSSSSS
    6    .FRRRRRRRRREFFFFEFFFFFIFFFFIFFFFFFMMMMMMMMMMSSSSSSSSSSSSSSSSSSSSS   SSSSSS
    5    .FRRRRRRRRRFFFFFFFFFFFFFFFFFFFFFFFFFMMMMMMMMMSSSSSSSSSSSSSSSSSSSS   SSSSSS
    4    .FRRRRRRRRREFFFFEFFFFFIFFFFIFFFFFFMMMMMMMMMMSSSSSSSSSSSSSSSSSSSSS   SSSSSS
    3    .FRRRRRRRRRFFFFFFFFFFFFFFFFFFFFFFFFFFMMMMMMMMMSSSSSSSSSSSSSSSSSSS   SSSSSS
    2    .FRRRRRRRRREFFFFEFFFFFIFFFFIFFFFFFFFFMMMMMMMMMSSSSSSSSSSSSSSSSSSS   SSSSSS
    1    .................................................................  .....
         12345678901234567890123456789012345678901345678901234567890123451  23456
                 10        20        30        40        50        60          146
NOTE: R-RESIDENTIAL AREA, E-CANDIDATE EXTRACTION, I- CANDIDATE INJECTION, E-OPTIMAL EXTRACTION, I-OPTIMAL INJECTION
```

Figure 6.25: The SW toe and FW tip simulated with the twentieth realization of the hydraulic conductivity after the third iteration.

6.4. Conclusions

In this chapter, how the deterministic and uncertainty problems have been applied to the hypothetical model has been examined. Solving the uncertainty problem is rather time-consuming if compared to the corresponding deterministic problem. The time needed for a one-iteration run of the uncertainty case (with 20 realizations) is recorded as seven times larger than for the one of the deterministic case.

By this hypothetical application, it is possible to make the conclusions:

- The symmetry of the hypothetical model helps to check the validity of the multi-objective optimization programs in the cases where the deterministic problems (for both methods) give the symmetric optimal solutions and the simulated saltwater intrusion increments are also symmetric.

-The uncertainty of the aquifer transmissivities makes the interfaces closer to the capture zone of the same optimal wells if compared to the deterministic case. This also implies that the risk due to the uncertainty of the aquifer parameter is taken into the computation. Hence, the average saltwater intrusion increment predicted is greater than in the deterministic case where there is no risk of the parameter uncertainty.

- This will help to realize that the saltwater intrusion management of the groundwater aquifer under the uncertainty always incurs more costs in order to achieve the same level of saltwater intrusion control.

- Regarding the awareness of the uncertainty risk, the uncertainty problem will trade off the time consumption against the safety of the optimal solution and, hence, the implementation of the optimized management scheme.

- The results of the multi-objective optimization problems solved by the two methods show the trade-off among the optimal values (Z_1, Z_2) that are obtained in the weighting method (with given w-weight) and in the L_2-metric method. This also shows the flexibility of the weighted problem in adjusting the saltwater intrusion level in the multi-objective management problem by only altering the weight value.

- The small differences between the optimal values, Z_2, and the simulated saltwater intrusion increments for most cases help to certify the accuracy of the sequential linearization approach that has been used in the programs.

- Generally, we can split up the non-inferior set in two sets such that for points in the first set the MRT is always bigger than for points in the second set. It reflects the fact that application of higher costs for injection is after some point (Z_1, Z_2) less effective.

Chapter 7

The Results of The Real-World Case Application

7.1 Introduction

The real-world problem for the proposed methodology is addressed in one particular study area that has been selected from the coastal areas in the Mekong Delta in Viet Nam. The area is intruded by saltwater with the current interface position located near the pumping wells. The two-dimensional shape of this modelled area is shown in Figure 7.1. For this particular area, the available data are very scarce and only the averaged values of all aquifer properties are given, e.g. hydraulic conductivity, thickness and cross sectional geometry. Unlike the modelled area in the hypothetical case, in the real-world case the modelled area has an asymmetric shape in the horizontal plane with its boundary based on the rivers surrounding it (see Figure 7.1). For the model boundary the fixed-flux boundary is set such that the current salt/freshwater interface position is maintained. The aquifer system is underlain by an uppermost layer with a low permeability (c = 5780 days), which is called a semi-permeable, or confining, layer. The lower layer has higher permeability (more than 500 times as large), which is called the aquifer. This aquifer is partly intruded by saltwater, as previously stated. The aquifer is considered to be confined, and having roughly the same thickness in the cross-sectional profile.

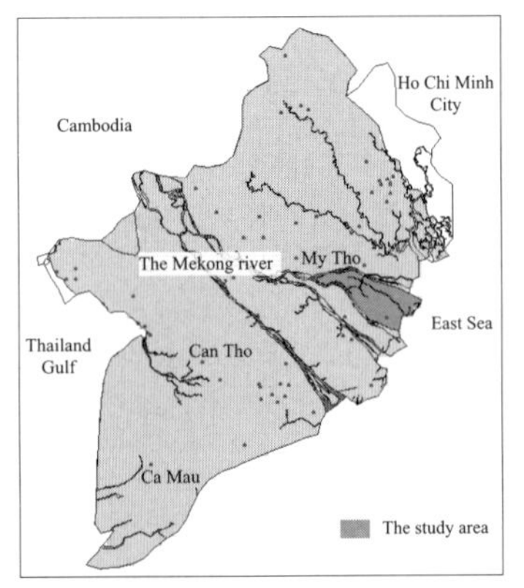

Figure 7.1: The location of the study area in the Mekong Delta.

In the horizontal plane, the two-dimensional grid system, in which there are 29 rows of 1250m each in the y-direction, is simulated. Grid spacing in the x-direction is 1250m except near the location of wells and the interface where it is reduced to 250m in order to improve the interface projection accuracy. In the model, there are 150 columns with a length of 250m and 36 columns with a length of 1250m. Therefore, in the x-direction the total number of columns is 186 with a total length of 82500m, which is equivalent to 66 columns × 1250m. The SW intrusion length is taken as the average of the saltwater toe intrusion lengths of all rows in the horizontal plane. The salt/fresh interface is considered as an abrupt change of the groundwater density (from 1026 mg/l to 1000mg/l). In the model this interface is roughly set to coincide with the

1‰ isohaline that bounds the current saltwater zone including the mixing zone of brackish water.

The hydro-geological data and aquifer parameters in the deterministic problem will be as follows:

The topmost layer is called the semi-pervious layer (aquitard), with a leakance value of about 1.73 10^{-4} /day (≈ 2.0 10^{-9} /s).

For the deterministic case, it is assumed that the permeable layer (main aquifer) is homogenous with its hydraulic conductivity of 5 m/day (≈ 5.787 10^{-5} m/s) and its thickness of 151m.

The bottom of the system is considered to be impermeable, as it has a no-flow boundary.

A constant flux boundary is put on its boundary. Its values are set in the range of 360 - 1800 m^3/h (≈ 0.1 - 0.5 m^3/s) so that the current SW interface is maintained.

The plan-view geometry of the model is illustrated in Figure 7.2.

The initial interface at the non-pumping condition for this problem is simulated and mapped in Figure 7.3.

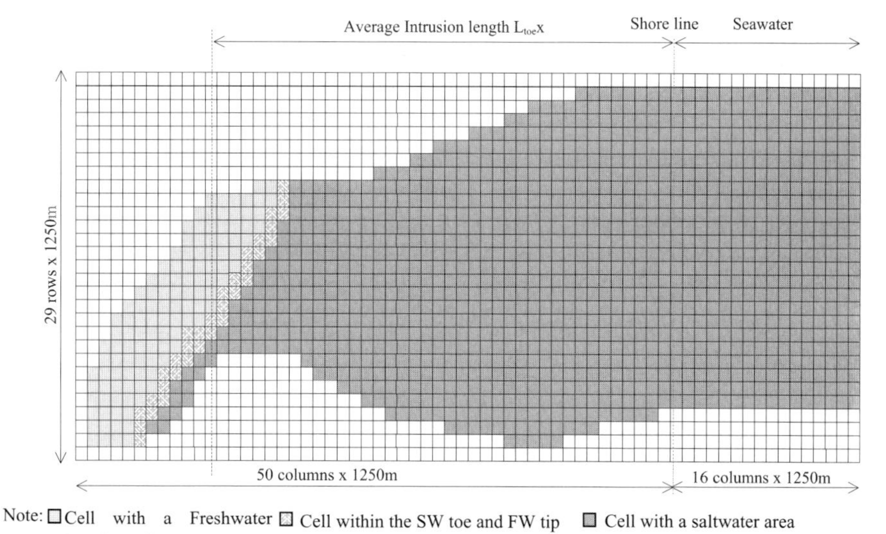

Figure 7.2: The plan view of the study area (with 66 columns × 1250m).

For the stochastic optimization problem, the uncertainty involved in the problem is mainly considered by the uncertain values of the hydraulic conductivity distributed in the discretized model. The distributed hydraulic conductivity values are randomly generated within the range from 4.5 m/day to 35 m/day with the kriging variance, cc = 1.0. The other aquifer parameters are kept the same as in the deterministic case.

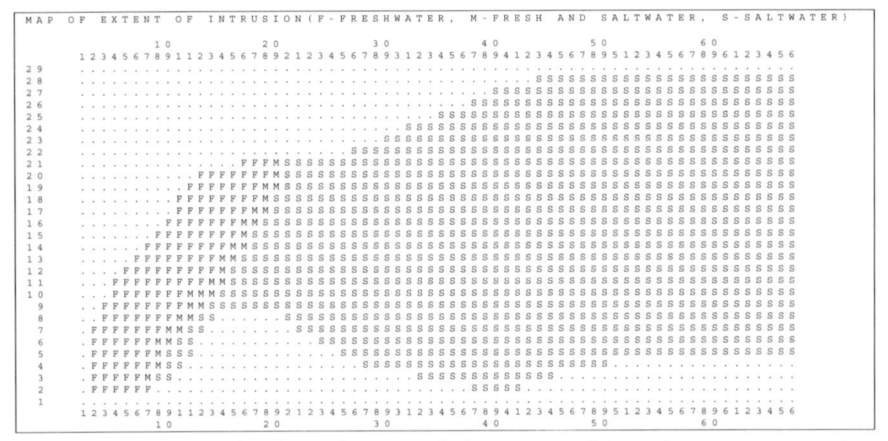

Figure 7.3: The simulated result map of the extent of intrusion at non-pumping condition (with 66 columns × 1250m).

7.2 The SW intrusion planning and management for the real-world problems

This management scheme with the prevention of saltwater intrusion by artificial injection, as proposed here, is new and can be applied in areas where there is a potential risk of saltwater intrusion. Therefore, the saltwater intrusion management problem presented in this work will give more insight in the sense of planning rather than the current management scheme of the study area.

Since the study area is subject to the tropical monsoon weather, there are two distinct seasons, the wet and the dry seasons. The wet and dry seasons take place alternately during six-month periods every year. During the wet season from May to November there is a periodical flood that makes the discharge of the Mekong River system much higher than the low-flow discharge in the dry season. It creates a need to regulate the flow of the river, so as to evacuate the floodwater on the ground surface in the wet season and increase the water supply for the dry season water demand. Under these circumstances, a scheme of management is proposed – a so-called seasonal planning for the saltwater intrusion management problem for this particular region. This consists of two managerial stages during each year – the first stage for the saltwater intrusion management during the wet season and the second stage during the dry season. The programs will run for the first stage and its solution, which is dependent on the L_α-metric or weight values of the two objectives, will be attained. The salt/fresh interface that is simulated using the model with the optimal solution obtained in the first stage will be the initial interface for the second stage of the management problem.

7.2.1 The first stage for the SW management during the wet season

For this stage the extraction and injection pumping will be operated simultaneously. The purpose is to store as much freshwater from the river in the flood season into the aquifer as possible. In this season, the supply water demand for domestic use from extraction wells is not as high as the water storage requirement in the aquifer by the injection wells. At the end of this stage the freshwater zone should be extended as far as possible. How far that is depends on the compromised solution from the multi-objective problem, which tends towards the objective of moving the SW interface seaward. The optimization problem for the first stage is operated roughly within six months in the wet season and so is the simulation model SHARP of the SW interface.

7.2.2 The second stage for the SW management during the dry season

At the end of the first stage the position of the SW interface is used for the input of the second-stage management problem. Then the SHARP simulation program will resume the continuation run with the updated input data from the results of the first-stage simulation program. The optimization program in this second stage is comprised of the decision variables for the extraction wells only. This is because in the dry season the river water near the estuaries is also polluted by seawater intrusion. Consequently, the scarce fresh water in the river is used for drinking water supply as the first priority. Hence, no fresh water at all, or at best only a very small amount of this freshwater, can be used for injecting in the aquifer. While the water supply demand in this period is the highest throughout the year, the extraction rate of groundwater can be increased to meet that demand. However, either an increase in, or constancy of, the extraction rate without aquifer supplementation by injection will cause the SW intrusion through the land-ward movement of the SW interface during the dry season operation. Therefore, the difference from the first stage is to keep the injection at zero and to double the extraction demand within six months. Besides that, the second-stage optimization problem will differ from the first-stage problem by setting the positive right-hand side for the SW intrusion constraint. This is because $d = L_{max} - L_0 > 0$, where L_{max}, which is the maximum SW intrusion length allowed in the dry season, is bigger than L_0 which is the initial SW intrusion length at the beginning of the second stage or at the end of the first stage.

For the real-world case problem the two-stage management scheme to both the deterministic case and uncertainty case problems will also be applied. Because of the flexibility of the weighting method, therefore, the weighting method will be preferable for the first stage of these two-stage management problems. The advantage of this method is that the interface can be regulated at the end of the first stage by varying the weight value, *w*. This means that, at the end of the second stage, the interface cannot pass the boundary where the initial interface was set at the beginning of the first stage.

7.3 The formulation of the real-world case's optimization problems

7.3.1 Decision variables

In the general optimization problem, the candidate wells (the number of decision variables) could possibly be located at all cells of the area where the freshwater zone exists in its aquifer. Besides, the existing wells can also be the candidate wells if one would like to evaluate the efficiency of those installed wells in the optimal sense. The optimization problem with more wells (and consequently, more variables) will require more time to solve it. In order to save computing time for this application problem, however, the candidate wells will be selected with the following distribution rule: -the injection wells are closer to the salt/fresh water interface toe whereas the extraction wells are far away from that interface. In this problem, the 40 candidate wells, i.e. 20 partly-penetrating extraction wells and 20 fully-penetrating injection wells, are located at rows 2 to 21 which correspond to the columns in Table 7.1 and as shown in Figure 7.4.

7.3.2 Objectives:

The multi-objective programming problems are formulated under the set of 40 decision variables (N_w = 40) that represent 20 candidate extraction wells and 20 candidate injection wells.

Two objectives will be considered i.e.

- The minimization of the distance between the location of the interface toe and the shoreline at the end of the simulation time, which is in either the first or second stage. This objective is equivalent to the minimization of the saltwater intrusion increment within that simulation period, as follows:

$$\underset{x}{\text{Min}}\ [Z_2 = \lambda^T x],$$

where λ is the response matrix of the saltwater intrusion increment and x is the vector of decision variables.

The above formula is only used for the deterministic case, whereas for the uncertainty case this objective will be formulated in a different form, such as:

$$\text{Min}\ \ \bar{\lambda}^T x + \left\| P_0^T x \right\|_2,$$

where λ is replaced by the nominal values, $\bar{\lambda}$, and the perturbation matrix, P_0, in the objective.

- The minimization of total costs due to extraction and injection strategies.

$$\underset{x}{\text{Min}}\ [Z_1 = c^T x],$$

where c^T is the vector of coefficients that represent the pumping and installation costs for extraction wells, c_{ext}, and injection wells, c_{inj}. The costs are transferred to the amount of monetary units paid for every cubic meter of water in the problem. Here it is assumed that these costs are set constants so that this objective is a linear function with respect to x (the design vector of 40 decision variables of pumping rates). Therefore the vector c^T = $[[c_{ext}]^T, [c_{inj}]^T]$ where c_{ext} = 0.05 MU/m^3 for each of 20 extraction wells and c_{inj} = 0.1 MU/m^3 for each of 20 injection wells. The costs for the injection are considered higher than for the extraction because the treatment of river water must be carried out before the injection and the fully penetrating injection wells must carry a higher price for their installation.

However, for the second-stage management, the variables that represent the injection rates of the candidate injection wells all have zero values. This makes the non-inferior set become a vertical line for any iteration. It is because when the cost objective is minimized in such a way that the water demand is satisfied and without injection that the optimal value is always equal to a constant pumping cost for that minimal water demand. This optimal value is always unique even though it is not necessary for the optimal solution to be unique for most of its iterations. It is also because the different costs (unit price for 1m^3 water × distance) due to the different distances of wells from a referent location are not taken into account.

For the first stage, those two objectives are set such that they are not only conflicting (decreasing the intrusion length will necessarily require an increase of the injection rate and consequently the operational cost) but also non-commensurate (different units in the objectives attributes). Therefore the solution process is quite complex.

However, in the second stage the two objectives are not always conflicting. This is because the operational cost objective is always constant, regardless of the slight variation of the intrusion length.

7.3.3 Constraints:

- ***Surface water injection rate upper bounds***

For the wet season, 0.25 m^3/s (≈ 900 m^3/h) is set as the upper bound of all injection wells at one cell (equivalent to 10 wells/cell x $90 m^3 h^{-1}$/well). Though the freshwater source (from the river) for the injection could be another limitation, here it is assumed that the river discharge in the wet season is high enough to be unlimited for injecting. For the dry season, because of the low flow of the rivers, the injection wells are totally switched off,

$$x_j \le M_j^{inj} \begin{cases} M_j^{inj} = 900 & \text{for the wet season} \\ M_j^{inj} = 0 & \text{for the dry season} \end{cases}, \qquad j = 1,\ldots, N_{iw}$$

- ***Extraction well capacity constraints***

Pumping rates from potential extraction cells are limited to a maximum yield of 50 m^3/h (≈ 0.014 m^3/s) for the whole year round.

$$x_j \le M_j^{ext} = 50, \qquad j = 1,\ldots, N_{ew}$$

- ***Supplying demand constraints***

The minimum water demand, W_D, for the region is of about 360 m^3/h (= 0.1 m^3/s) in the wet season and 720 m3/h in the dry season. This demand is only for domestic use.

$$\sum_{j=1}^{N_{ew}} x_j = e^T x \ge W_D \begin{cases} W_D = 360 & \text{for the wet season} \\ W_D = 720 & \text{for the dry season} \end{cases}.$$

- ***Draw-down constraints***

The maximum draw-down of piezometric head in the freshwater zone is not allowed to be greater than 10m because of the prevention of subsidence in the residential area. Hence, $b = -10$. In the residential area, it is necessary to set only the 24 control points ($N_c = 24$) that are located in the boundary cells (see Figure 7.4) that can guarantee the higher heads for the rest of cells of the whole residential area. Thus there will be 24 drawdown constraints for the deterministic problem under the forms as:

$$\sum_{j=1}^{N_{ew}+N_{iw}} A_{ij} x_j \ge b_i = -10, \; i = 1,\ldots, N_c = 24; \; j = 1,\ldots, N_w = 40,$$

where A_{ij} is the head response matrix.

For the uncertainty problems these constraints will be reformulated as

$$\bar{a}_i^T x - \left\| P_i^T x \right\|_2 \ge b_i, \qquad i = 1,\ldots,N_c$$

where A_{ij} are replaced by the nominal values, $\bar{a}_i$, and the perturbation matrices, P_i, in those constraints.

- ***Saltwater toe location constraint***

For the wet season, the maximum intrusion length is defined by the constraint of the increment of saltwater toe location. This is chosen such that the interface toe cannot reach either the well screen locations or the initial (at the beginning of pumping period) interface toe location during the operation. This is called the first option of SW intrusion management. The second option that will be selected for this constraint is to extend the freshwater zone during the operational process. This means that during the pumping period more freshwater will be stored in the aquifer than exploited. For the latter option, an average distance of 55km land-ward from the shoreline (100 cells away from the shoreline) is the upper bound of the intrusion length, L_{max}. The average distance of the initial interface toe from the shoreline, L_0, is set at about 55286 m. Then the average SW intrusion increment, $d = L_{max} - L_0$, is about –286m and, if chosen as the first option, then it can be seen that $d = 0$.

For the dry season (second stage), the interface that is simulated after the wet season (first stage) will be set as the initial interface for the second stage of management problems. The maximum saltwater intrusion length can then be set such that its corresponding interface roughly coincides with the initial interface at the beginning of the first stage. This means that L_{max} = 55286 m and $d = L_{max} - L_0 > 0$. If the maximum saltwater intrusion length, L_{max}, is expected to be some distance less than that value (55286 m) then d is less positive and after the one-year management the aquifer will gain at least such a distance of freshwater. Obviously in order to obtain such a solution the problem must be feasible, since under the effects of the groundwater extraction and the salt groundwater flow from the sea the interface will recede into the position where the intrusion length measured from the shoreline must be smaller than or equal to L_{max}.

$$\sum_{j=1}^{N_{ew}+N_{iw}} \lambda_j x_j \le d \begin{cases} d \le 0: & \text{for the wet season,} \\ d > 0: & \text{for the dry season.} \end{cases}$$

For the uncertainty case, the above linear inequality can be transformed into the non-linear inequality, as

$$\bar{\lambda}^T x + \left\| P_0^T x \right\|_2 \le d \begin{cases} d \le 0: & \text{for the wet season,} \\ d > 0: & \text{for the dry season,} \end{cases}$$

where λ are replaced by the nominal value, $\bar{\lambda}$, and the perturbation matrix, P_0.

- ***Non-negative constraints***

The non-negativity can be imposed on the pumping rates of the candidate wells by the non-negativity constraints as:

$$x_j \ge 0, \qquad j = 1,\dots,N_{iw}+N_{ew}$$

This type of constraint guarantees that the optimal solution will always be positive. Consequently, in the optimal solution vector very small values (less than 10^{-3} m^3/h) for its components may occur that can be neglected and reported as equal to zero.

In brief, for the deterministic multi-objective SW intrusion management problems there are the following formulas:

$$\underset{x}{\text{Min}} \quad [Z_1 = c^T x], \tag{7.1}$$

$$\underset{x}{\text{Min}} \quad [Z_2 = \lambda^T x], \tag{7.2}$$

subject to

$$\sum_{j=1}^{N_{ew}+N_{iw}} A_{ij} x_j \geq b_i \ , \qquad i = 1,\ldots,N_c \tag{7.3}$$

$$\sum_{j=1}^{N_{ew}+N_{iw}} \lambda_j x_j \leq d \begin{cases} d \leq 0: \text{ for the wet season} \\ d > 0: \text{ for the dry season} \end{cases}, \tag{7.4}$$

$$\sum_{j=1}^{N_{ew}} x_j \geq W_D \begin{cases} W_D = 360 \quad \text{for the wet season} \\ W_D = 720 \quad \text{for the dry season} \end{cases}, \tag{7.5}$$

$$x_j \leq M_j^{inj} \begin{cases} M_j^{inj} = 900: \text{ for the wet season} \\ M_j^{inj} = 0 \quad : \text{ for the dry season} \end{cases}, j = 1,\ldots, N_{iw} \tag{7.6}$$

$$x_j \leq M_j^{ext} \ \{ M_j^{ext} = 50 \ , \qquad j = 1,\ldots, N_{ew} \tag{7.7}$$

$$x_j \geq 0 \, , j = 1,\ldots,N_{iw}+N_{ew} \, , \tag{7.8}$$

where $c^T = [0.05,\ 0.1]^T$, $b_i = -10$, λ_j and A_{ij} are computed by the SHARP and PERTURB computer codes.

In addition, for the uncertainty multi-objective SW intrusion management problems there are the following formulas:

$$\underset{x}{\text{Min}} \quad [Z_1 = c^T x], \tag{7.1}$$

$$\underset{x}{\text{Min}} \quad \bar{\lambda}^T x + \left\| P_0^T x \right\|_2 , \tag{7.2'}$$

subject to

$$\bar{a}_i^T x - \left\| P_i^T x \right\|_2 \geq b_i \ , \qquad i = 1,\ldots,N_c \tag{7.3'}$$

$$\bar{\lambda}^T x + \left\| P_0^T x \right\|_2 \leq d \begin{cases} d \leq 0: \text{ for the wet season} \\ d > 0: \text{ for the dry season} \end{cases}, \tag{7.4'}$$

$$\sum_{j=1}^{N_{ew}} x_j \geq W_D \begin{cases} W_D = 360 \quad \text{for the wet season} \\ W_D = 720 \quad \text{for the dry season} \end{cases}, \tag{7.5}$$

$$x_j \le M_j^{inj} \quad \begin{cases} M_j^{inj} = 900: \text{ for the wet season} \\ M_j^{inj} = 0 \quad : \text{ for the dry season} \end{cases}, j = 1,..., N_{iw} \qquad (7.6)$$

$$x_j \le M_j^{ext} \quad \{M_j^{ext} = 50 \,, \qquad j = 1,..., N_{ew} \qquad (7.7)$$

$$x_j \ge 0 \,, j = 1,...,N_{iw}+N_{ew} \,, \qquad (7.8)$$

where $c^T = [0.05, 0.1]^T$, $b_i = -10$. The vectors, $\bar{\lambda}$, $\bar{a}_i$ are the nominal values and P_0, P_i are the perturbation matrices. Please refer to Chapter 5 for the computation.

7.4 The deterministic case

For the deterministic case, it is assumed that the aquifer is homogenous with the constant hydraulic conductivity value of 5 m/d. Figure 7.4 shows one modelled-area's part that has 100 columns of 250-meter size. In this problem, the 40 candidate wells, i.e. partly-penetrating extraction and fully-penetrating injection wells, are located at rows 2 to 21 which correspond to the columns in Table 7.1. Their locations are shown in Figure 7.4.

Table 7.1: The locations (row, column) of the candidate wells

Well Number	Extraction wells		Well Number	Injection wells	
	Row	Column		Row	Column
1	3	9	21	2	16
2	5	16	22	4	19
3	7	19	23	6	24
4	9	23	24	8	30
5	11	28	25	10	33
6	13	38	26	12	41
7	15	43	27	14	45
8	17	63	28	16	58
9	19	70	29	18	74
10	21	73	30	20	78
11	2	10	31	3	20
12	4	13	32	5	27
13	6	18	33	7	30
14	8	16	34	9	34
15	10	27	35	11	39
16	12	35	36	13	49
17	14	39	37	15	54
18	16	53	38	17	74
19	18	68	39	19	81
20	20	72	40	21	84

7.4.1 The saltwater intrusion management in the wet season (the first stage)

In Figure 7.4, it can be seen that the initial interfaces (toe and tip) that are simulated at the non-pumping condition are used for the input file of the simulation model for the first-stage management. The candidate wells are all located in the fresh water zone adjacent to the interface toe.

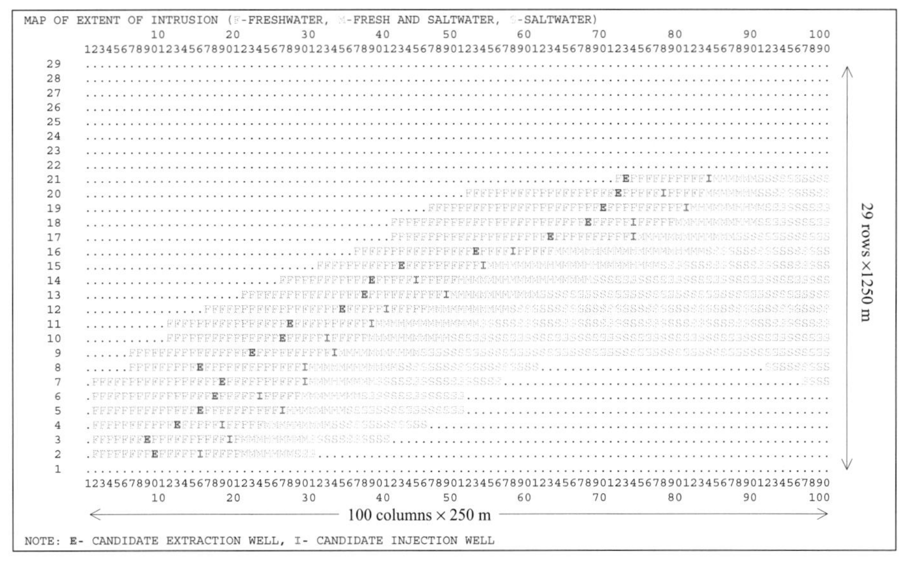

Figure 7.4: The interface toe and tip in the modelled area in the case of non-pumping condition (k = 5 m/d).

- ***The results of the multi-objective optimal solution by L_2-metric approach***

1. The optimal solution $x = \boldsymbol{Q}_{opt}$, at all iterations are shown in Table 7.2. The problem converges at the third iteration. The difference between the optimal solutions of the first and last iterations takes place on the pumping rates of the injection well 34, e.g. 865.14 m^3/h is changed into 900 m^3/h when ΔQ is initially equal to 10 m^3/h and gradually decreased to 9.6 m^3/h.

2. The optimal (trade-off) values of Z_1 and Z_2 and the SW intrusion increment, *Lsim*, which are shown in Table 7.3, are based on the optimal solution of the L_2-metric problem at different iterations. The non-inferior sets drawn in Figures 7.5 and 7.6 are created by the constraint method with the response matrices $[\Delta H_i / \Delta Q_j]$ and $[\Delta L / \Delta Q_j]$ that are based on the optimal solution from the L_2-metric problem. The optimal values of the L_1-metric and L_{inf}-metric computed are based on the same response matrices generated from the L_2-metric problem. Those values of the two problems are just for showing the narrow range of choice in the non-inferior set for the L_α-metric problems when α varies from 1 to infinity. This procedure of generating the non-inferior set and computing the optimal values of the L_1-metric and L_{inf}-metric is applicable for all the L_2-metric problems in this chapter.

At the last iteration, the optimal value Z_2 (−1910.66) is greater (1.7%) than the simulated saltwater intrusion increment *Lsim* (−1942.49) or Z_2 > *Lsim*. The trade-off values of (Z_1, Z_2) in the non-inferior set at the first and last iterations are depicted in Figures 7.5 and 7.6. It can be seen that the L_1-metric problems always give compromised solutions, which are mostly in favour of the Z_2 objective among the L_α-metric problems.

3. The optimal solution, $x = \boldsymbol{Q}_{opt}$, is distributed to 18 active wells, i.e. 8 extraction wells and 10 injection wells as shown in Table 7.4. This optimal solution is obtained at the third iteration (k = 3). The reader can refer these values of $\boldsymbol{Q}_{opt}$ to Figure 7.8

that is the output file of SHARP for mapping the extent of SW intrusion based on the last-iteration optimal solution vector.

4. The SW intrusion interface simulated by SHARP computer code

The map in Figure 7.7 shows the initial interface when there is no stress applied. With the optimal solution achieved by the L_2-metric optimization program, the interface moves seaward as shown in Figure 7.7. This map results from the SHARP simulation model.

Discussion

It can be seen in Figure 7.7 that most of the extraction wells with the highest rates are optimally allocated in the southernmost part of the modelled area, whereas most of the injection wells are in the middle and the north part. This shows the vulnerability of the middle and the north part with respect to the saltwater intrusion.

This can be explained by the fact that the saltwater intrusion response coefficients of most extraction wells that are located in the southernmost part are the smallest values (always positive). At the same time, the response coefficients of most injection wells that are located in the middle and north part are the smallest values (always negative). These smallest coefficients of the extraction wells in the southernmost area are caused by the higher boundary-flow assigned for that particular area.

Table 7.2: The results of the L_2-metric problem under deterministic case.

	Iteration number, k		
	1	2	3
ΔQ (m^3/h)	10.00	9.80	9.60
Z_1 (MU/h)	914.514	918.000	918.000
Z_2 (m)	-2676.807	-1885.997	-1910.659
Lsim	-1910.868	-1942.501	-1942.488
$\lVert(Q_i^{\,k} - Q_i^{\,k-1})\rVert$	10000.000	34.859	0.000
Q_1 (extraction well)	50.000	50.000	50.000
Q_2 (extraction well)	50.000	50.000	50.000
Q_3 (extraction well)	50.000	50.000	50.000
Q_4 (extraction well)	0.000	0.000	0.000
Q_5 (extraction well)	0.000	0.000	0.000
Q_6 (extraction well)	0.000	0.000	0.000
Q_7 (extraction well)	0.000	0.000	0.000
Q_8 (extraction well)	0.000	0.000	0.000
Q_9 (extraction well)	0.000	0.000	0.000
Q_{10} (extraction well)	10.000	10.000	10.000
Q_{11}(extraction well)	50.000	50.000	50.000
Q_{12} (extraction well)	0.000	0.000	0.000
Q_{13} (extraction well)	50.000	50.000	50.000
Q_{14} (extraction well)	50.000	50.000	50.000
Q_{15} (extraction well)	0.000	0.000	0.000
Q_{16} (extraction well)	0.000	0.000	0.000
Q_{17} (extraction well)	50.000	50.000	50.000
Q_{18} (extraction well)	0.000	0.000	0.000
Q_{19} (extraction well)	0.000	0.000	0.000
Q_{20} (extraction well)	0.000	0.000	0.000
Q_{21} (injection well)	0.000	0.000	0.000
Q_{22} (injection well)	0.000	0.000	0.000
Q_{23} (injection well)	0.000	0.000	0.000
Q_{24} (injection well)	900.000	900.000	900.000
Q_{25} (injection well)	0.000	0.000	0.000
Q_{26} (injection well)	900.000	900.000	900.000
Q_{27} (injection well)	0.000	0.000	0.000
Q_{28} (injection well)	900.000	900.000	900.000
Q_{29} (injection well)	900.000	900.000	900.000
Q_{30} (injection well)	0.000	0.000	0.000
Q_{31} (injection well)	900.000	900.000	900.000
Q_{32} (injection well)	0.000	0.000	0.000
Q_{33} (injection well)	0.000	0.000	0.000
Q_{34} (injection well)	865.141	900.000	900.000
Q_{35} (injection well)	900.000	900.000	900.000
Q_{36} (injection well)	0.000	0.000	0.000
Q_{37} (injection well)	900.000	900.000	900.000
Q_{38} (injection well)	900.000	900.000	900.000
Q_{39} (injection well)	900.000	900.000	900.000
Q_{40} (injection well)	0.000	0.000	0.000

Table 7.3: The optimal values and the *Lsim* at all different iterations (L_2-metric).

Iteration	ΔQ (m³/h)	Z_1 (MU/h)	Z_2 (m)	*Lsim* (m)	$\|(Q_i^{\,k} - Q_i^{\,k-1})\|$
1	10.00	914.51	−2676.80	−1910.86	10000.00
2	9.80	918.00	−1885.99	−1942.50	34.85
3	9.60	918.00	−1910.65	−1942.48	0.00

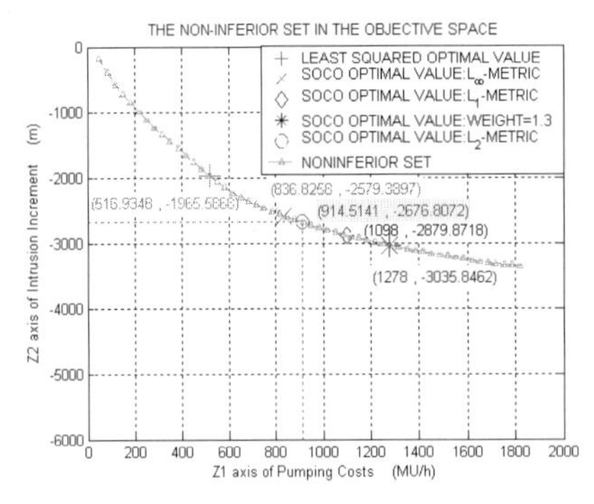

Figure 7.5: The first iteration.

Figure 7.6: The third iteration.

Table 7.4: The optimal solution at the last iteration (L_2-metric)

Well Number	Extraction wells Row	Column	Q_{opt} (m³/h)	Well Number	Injection wells Row	Column	Q_{opt} (m³/h)
1	3	9	50.000	21	2	16	0.000
2	5	16	50.000	22	4	19	0.000
3	7	19	50.000	23	6	24	0.000
4	9	23	0.000	24	8	30	900.000
5	11	28	0.000	25	10	33	0.000
6	13	38	0.000	26	12	41	900.000
7	15	43	0.000	27	14	45	0.000
8	17	63	0.000	28	16	58	900.000
9	19	70	0.000	29	18	74	900.000
10	21	73	10.000	30	20	78	0.000
11	2	10	50.000	31	3	20	900.000
12	4	13	0.000	32	5	27	0.000
13	6	18	50.000	33	7	30	0.000
14	8	16	50.000	34	9	34	900.000
15	10	27	0.000	35	11	39	900.000
16	12	35	0.000	36	13	49	0.000
17	14	39	50.000	37	15	54	900.000
18	16	53	0.000	38	17	74	900.000
19	18	68	0.000	39	19	81	900.000
20	20	72	0.000	40	21	84	0.000

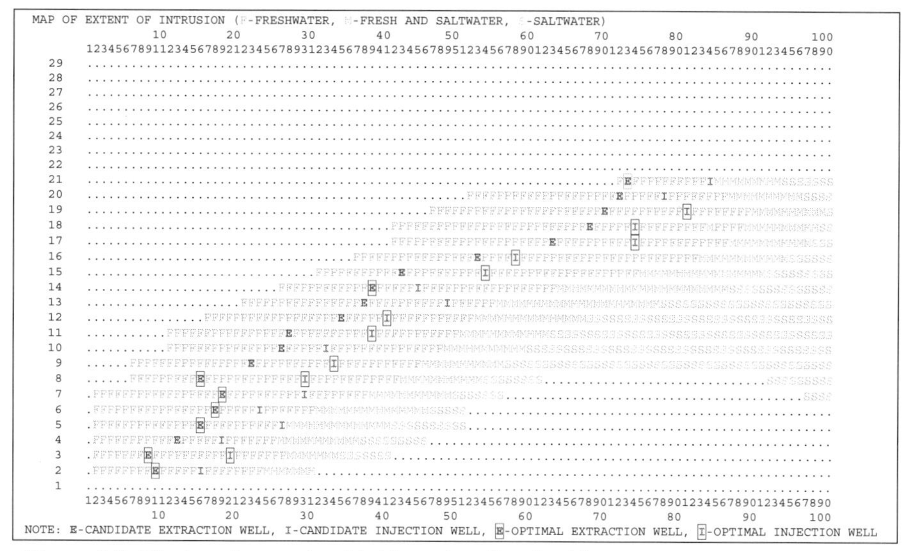

Figure 7.7: The interface at the third iteration ($k = 5$ m/d).

- ***The results of the multi-objective optimal solution by the weighted problem (w = 1.3)***

1. The results of this problem show the faster convergence of the optimal solution, $x = \boldsymbol{Q}_{opt}$. Table 7.5 shows that the optimal solution converges after two iterations.

2. The optimal values of Z_1 and Z_2, (1278, −2637.43), in this problem are more favourable to the Z_2 objective than in the L_2-metric problem. We can also see the trade-off between two problems by comparing the above optimal values with the optimal values of the L_2-metric problem, i.e. (918, −1910.66). The simulated SW intrusion increment, *Lsim*, i.e. −2531.37 m, which is shown in Table 7.6, is 4.3% larger than the optimal value Z_2 or $Z_2 < Lsim$.

The trade-off values of (Z_1, Z_2) in the non-inferior set at the first and last iterations are depicted in Figures 7.8 and 7.9. These non-inferior sets drawn in those figures are created by the constraint method with the response matrices [ΔH_i / ΔQ_j] and [ΔL / ΔQ_j]. These are based on the optimal solution from the $w = 1.3$ weighted problem. The optimal values with the weights $w = 0.1$ and $w = 5.0$ computed are based on the same response matrices generated from the $w = 1.3$ weighted problem. Those values of the two problems (with $w = 0.1$ and $w = 5.0$) in the non-inferior set are just for showing the full range of choice in the non-inferior set for the weighted problem. Such a procedure of generating the non-inferior set and computing the optimal values is applicable for all the weighted problems in this chapter.

3. The optimal solution, $x = \boldsymbol{Q}_{opt}$, indicates 22 wells to be allocated, i.e. eight extraction wells and 14 injection wells as shown in Table 7.7. In comparison with the L_2-metric problem, the optimal solution of the weighted problem ($w = 1.3$) is attained at four more injection wells, i.e. the wells numbers 22, 25, 33 and 36. At the same time the optimal extraction wells are kept the same in terms of the pumping rates and locations. The reader can refer these values of $\boldsymbol{Q}_{opt}$ to Figure 7.10, which is the output file of SHARP for mapping the extent of SW intrusion based on the last-iteration optimal solution vector.

4. The SW intrusion interface is simulated by SHARP computer code. In Figure 7.10 of the weighted problem, because of four more injection wells added to the middle south part, the interface in the southern area is closer to the shoreline than in Figure 7.7, which is the result map of the L_2-metric problem.

Discussion

If compared with the L_2-metric problem results, there are nine out of 14 injection wells of the weighted problem, which are more predominantly distributed in the middle area (i.e. in the ratio of nine to 14 as compared with the ratio of six to 10). The other three injection wells are allocated to the north, as in the L_2-metric problem, whereas one more injection well is added to the southernmost area for the weighted problem. The four additional injection wells with such a distribution result in the expansion of the freshwater zone in the southernmost area by driving back the interface seaward. With the greater values of w the weighted problem will give the optimal values that are more advantageous to the objective Z_2 and, consequently, the second stage of management problem will have more possibilities to be feasible in the sense of the saltwater intrusion.

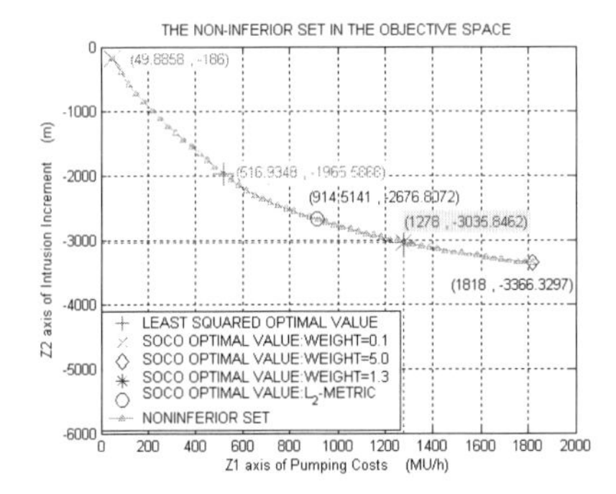

Figure 7.8: The first iteration.

Figure 7.9: The second iteration.

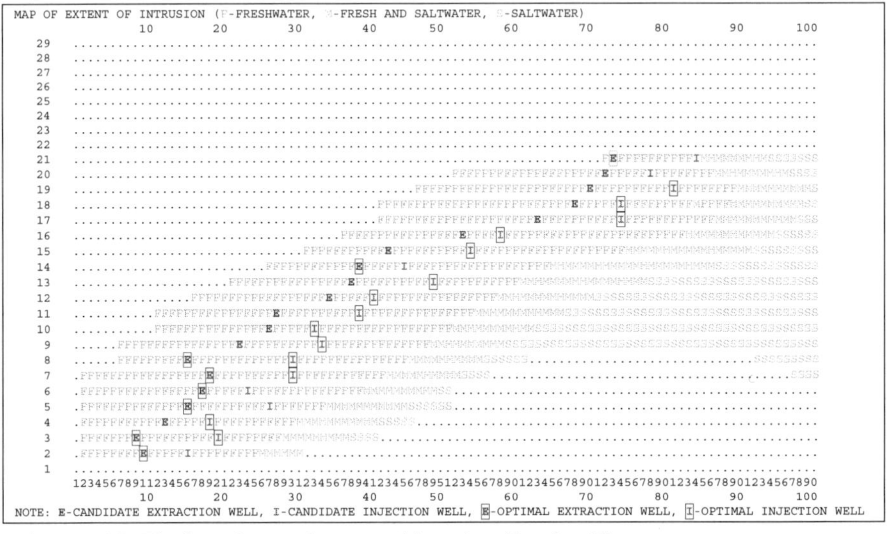

Figure 7.10: The interface at the second iteration (k = 5 m/d).

Table 7.5: The results of the weighted problem ($w = 1.3$) under deterministic case.

	Iteration number, k	
	1	2
ΔQ (m^3/h)	10.000	9.800
Z_1 (MU/h)	1278.000	1278.000
Z_2 (m)	−3035.846	−2644.691
Lsim	−2531.361	−2531.372
$\|(Q_i^{\,k} - Q_i^{\,k-1})\|$	10000.000	0.000
Q_1 (extraction well)	50.000	50.000
Q_2 (extraction well)	50.000	50.000
Q_3 (extraction well)	50.000	50.000
Q_4 (extraction well)	0.000	0.000
Q_5 (extraction well)	0.000	0.000
Q_6 (extraction well)	0.000	0.000
Q_7 (extraction well)	0.000	0.000
Q_8 (extraction well)	0.000	0.000
Q_9 (extraction well)	0.000	0.000
Q_{10} (extraction well)	10.000	10.000
Q_{11} (extraction well)	50.000	50.000
Q_{12} (extraction well)	0.000	0.000
Q_{13} (extraction well)	50.000	50.000
Q_{14} (extraction well)	50.000	50.000
Q_{15} (extraction well)	0.000	0.000
Q_{16} (extraction well)	0.000	0.000
Q_{17} (extraction well)	50.000	50.000
Q_{18} (extraction well)	0.000	0.000
Q_{19} (extraction well)	0.000	0.000
Q_{20} (extraction well)	0.000	0.000
Q_{21} (injection well)	0.000	0.000
Q_{22} (injection well)	900.000	900.000
Q_{23} (injection well)	0.000	0.000
Q_{24} (injection well)	900.000	900.000
Q_{25} (injection well)	900.000	900.000
Q_{26} (injection well)	900.000	900.000
Q_{27} (injection well)	0.000	0.000
Q_{28} (injection well)	900.000	900.000
Q_{29} (injection well)	900.000	900.000
Q_{30} (injection well)	0.000	0.000
Q_{31} (injection well)	900.000	900.000
Q_{32} (injection well)	0.000	0.000
Q_{33} (injection well)	900.000	900.000
Q_{34} (injection well)	900.000	900.000
Q_{35} (injection well)	900.000	900.000
Q_{36} (injection well)	900.000	900.000
Q_{37} (injection well)	900.000	900.000
Q_{38} (injection well)	900.000	900.000
Q_{39} (injection well)	900.000	900.000
Q_{40} (injection well)	0.000	0.000

Table 7.6: The optimal values and the *Lsim* at all different iterations (w = 1.3).

Iteration	ΔQ (m^3/h)	Z_1 (MU/h)	Z_2 (m)	*Lsim* (m)	$\|(Q_i^k - Q_i^{k-1})\|$
1	10.00	1278.00	–3035.84	–2531.36	10000.00
2	9.80	1278.00	–2644.69	–2531.37	0.00

Table 7.7: The optimal solution at the last iteration (w = 1.3)

Well	Extraction wells			Well	Injection wells		
Number	Row	Column	Q_{opt} (m^3/h)	Number	Row	Column	Q_{opt} (m^3/h)
1	3	9	50.000	21	2	16	0.000
2	5	16	50.000	22	4	19	900.000
3	7	19	50.000	23	6	24	0.000
4	9	23	0.000	24	8	30	900.000
5	11	28	0.000	25	10	33	900.000
6	13	38	0.000	26	12	41	900.000
7	15	43	0.000	27	14	45	0.000
8	17	63	0.000	28	16	58	900.000
9	19	70	0.000	29	18	74	900.000
10	21	73	10.000	30	20	78	0.000
11	2	10	50.000	31	3	20	900.000
12	4	13	0.000	32	5	27	0.000
13	6	18	50.000	33	7	30	900.000
14	8	16	50.000	34	9	34	900.000
15	10	27	0.000	35	11	39	900.000
16	12	35	0.000	36	13	49	900.000
17	14	39	50.000	37	15	54	900.000
18	16	53	0.000	38	17	74	900.000
19	18	68	0.000	39	19	81	900.000
20	20	72	0.000	40	21	84	0.000

7.4.2 The saltwater intrusion management in the dry season (the second stage)

Since the flexibility of the weighting method if compared to the L_α-metric method is known, the interfaces moved by varying the weight value, $w > 0$, in the weighted problems can be made. Consequently, the optimal values can be any extreme points in the whole non-inferior set; it is especially important for the points to the southeast of this non-inferior set (the points are in more favour of the objective Min Z_2). Therefore, in the second stage it is proposed that the saltwater intrusion management problem initializes the interfaces that result from the weighted problem in the first stage. Figure 7.13 shows the initial interface for the second stage; this interface is simulated with the optimal solution from the weighted problem (w = 1.3) in the first stage. The candidate wells are in the same location as in the first stage. The water demand in this period is doubled since more pumping water is required in the dry season.

In the second stage, the injection well capacity could be reduced or the upper bound constraint for the surface water that can be allowably used for injecting could be set. However, this case is still more or less the same as the first-stage management in the sense that firstly, both the extraction and injection exist, and secondly, they are performed simultaneously. Thus, this case is not considered in this thesis.

For the worst case in the second stage, e.g. all surface water sources are salt polluted, all the injection wells are switched off ($M_j^{inj} = 0$) and only the candidate extraction wells are taken into the optimizing process. Then the optimal solution is only distributed to the extraction wells and the summation of those rates equals the water demand for this period. Consequently, the objective Z_1 gets a fixed optimal value (i.e. $Z_1 = 36$ MU/h) that is the extraction costs for the minimal water demand (W_D is now set double, 720 m^3/h) without injecting costs. Therefore, the optimal values (Z_1, Z_2) of the second stage are of a fixed point of the non-inferior set in the objective space. In fact, the optimal solutions with respect to the two single objective problems can be different from each another. Hence, their optimal values (Z_1^{opt}, Z_2^{trad}) and (Z_1^{trad}, Z_2^{opt}) corresponding to Z_1 and Z_2 single objective problems have the co-ordinates in the pay-off table such that $Z_1^{trad} = Z_1^{opt}$ and $Z_2^{trad} > Z_2^{opt}$. Consequently, the non-inferior set could be a vertical line segment that has its coordinate at $Z_1 = 36$. In this non-inferior set, the utopia point (Z_1^{opt}, Z_2^{opt}) is located at the lowest point. The optimal values of the multi-objective problem (depending on α or w values) are always the co-ordinates of the higher point so that it could be very close to, but never coincide with, the utopia point when $\alpha > 1$ in the L_α-metric problems (or $0 < w < \infty$ in the weighted problems). This is because the non-inferior set in this second stage really has alternative optima. This means that all the solutions that lie along this vertical line segment are alternative optima except for the utopia point. This kind of non-inferior set happens since, firstly in the cost objective Z_1, the different costs due to different distances of well locations are not taken into account, and secondly, these two objectives are not always conflicting. Thus, in this second stage the utopia point is the best solution and one cannot find another point which is better. Therefore, the optimal solution for the multi-objective problem would be the optimal solution of either the L_1-metric problem or the $w = \infty$ weighted problem, if they exist. In the real-world case, the numerical results of the two application problems, i.e. the deterministic and uncertainty problems, will be observed in this chapter. The general demonstration is shown in Figures 7.11 and 7.12.

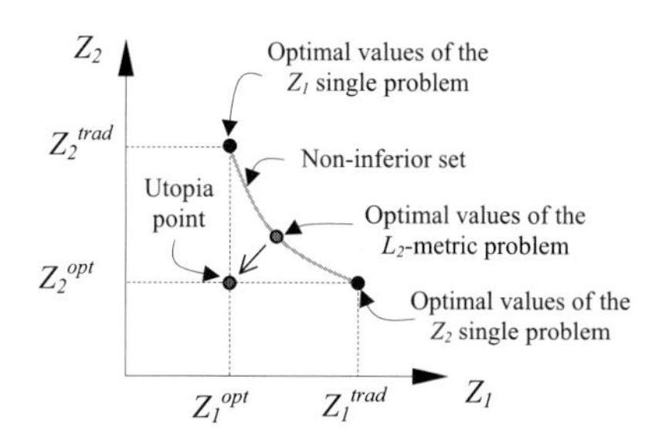

Figure 7.11: The non-inferior set for the first stage.

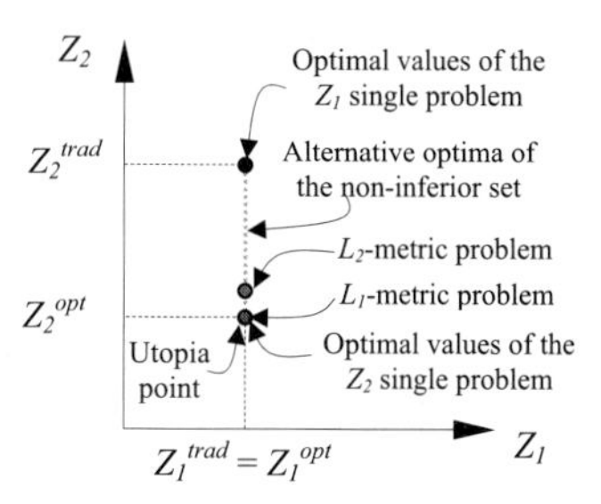

Figure 7.12: The non-inferior set for the second stage where the two objectives are not conflicting.

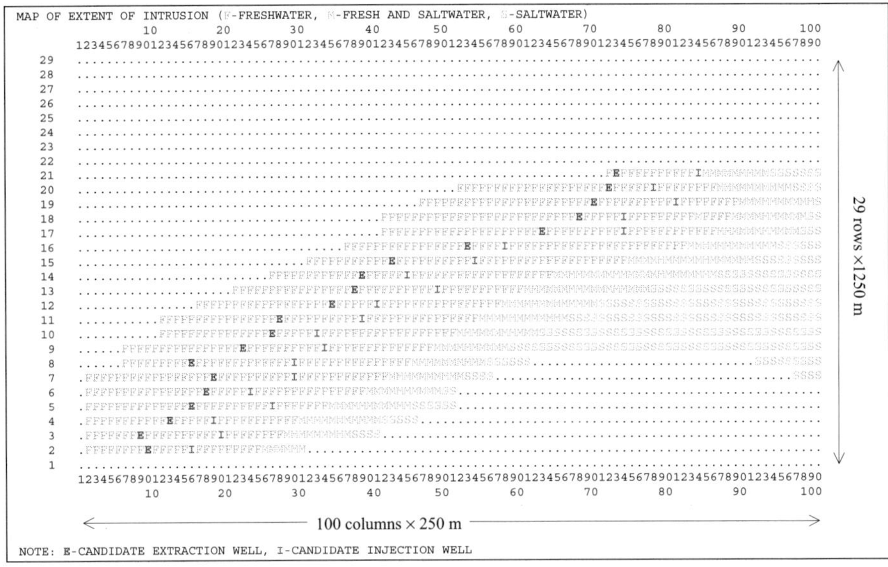

Figure 7.13: The initial interface for the second stage ($k = 5$ m/d).

- ***The results of the multi-objective optimal solution by the L_1-metric method***

1. The optimal solution converges after three iterations as shown in Table 7.8.

2. The optimal values of Z_1 and Z_2 and the SW intrusion increment, *Lsim*, which are shown in Table 7.9, are based on the optimal solution of the L_1-metric problem at different iterations. At the third iteration, the optimal value Z_2 is 1751.29 m, which indicates that the interface toe has moved land-ward. The simulated saltwater intrusion increment, *Lsim*, is also a positive value (1766.46 m) that approaches the optimal value Z_2, with $Z_2 < Lsim$. In this second stage, each iteration's optimal values are of only one point that coincides with the utopia point, as depicted in Figures 7.14 and 7.15.

3. The optimal solution $x = Q_{opt}$, which is shown in Table 7.10, is distributed to 15 active extraction wells. There are 14 injection wells with the maximum well capacity (50 m^3/h) that are mostly located in the south and middle parts. The remaining extraction well, which has the smallest rate of 20 m^3/h, is located in the north. The summation of these pumping rates is equal to the water demand (W_D = 720 m^3/h) that is set double for the dry season. These values of Q_{opt} are referred to in Figure 7.16, which is the output file of SHARP for mapping the extent of SW intrusion based on the last-iteration optimal solution vector.

4. The SW intrusion interface simulated by SHARP computer code

If Figure 7.13, which shows the initial interfaces at the beginning of the second stage, is compared with Figure 7.16, it can be seen that the saltwater toe approaches the well locations at the end of the second stage.

Discussion

If the result map (Figure 7.16) at the end of the second stage is compared with Figure 7.4, which shows the initial interface at the beginning of the first stage, it can be seen that their interfaces are nearly the same as each other. However, the interface at the end of the second stage is closer to the seashore than at the beginning of the first stage. Hence, the simulated freshwater zone is extended by an average distance of 764.91 m after a one-year management. This value is calculated by the difference between the simulated saltwater intrusion increments of the first and second stages (i.e. −2531.37 m and 1766.46 m, respectively). If calculated by the different optimal values at the end of the two stages (i.e. −2644.69 m and 1751.29 m, respectively), the average distance is about 893.40 m.

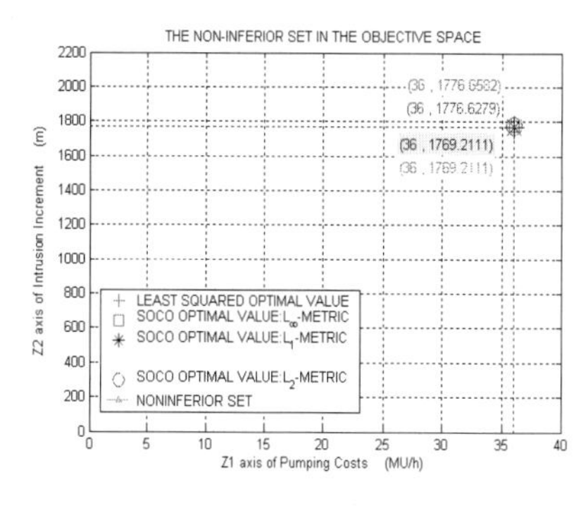

Figure 7.14: The first iteration.

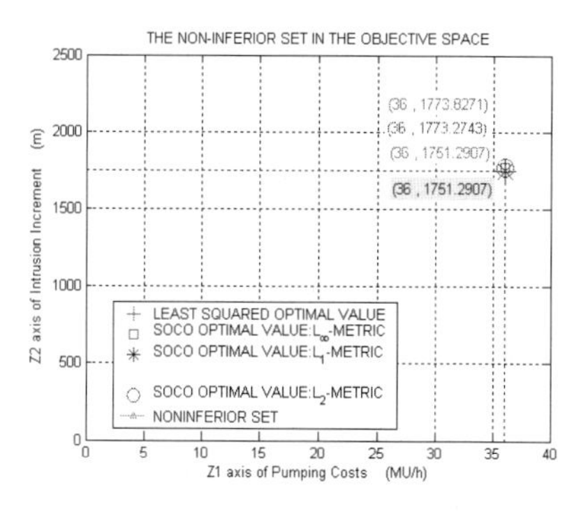

Figure 7.15: The third iteration.

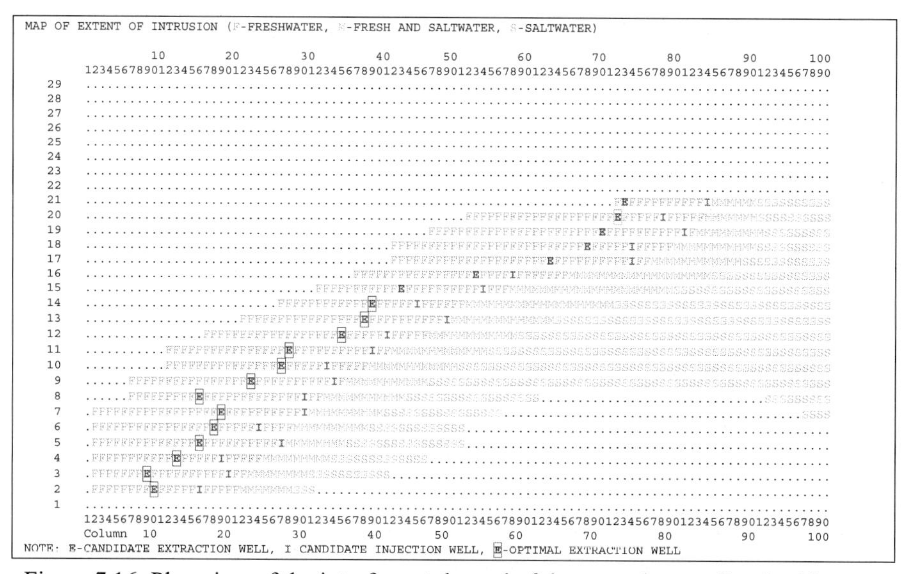

Figure 7.16: Plan view of the interface at the end of the second stage (k = 5 m/d).

Table 7.8: The results of the L_1-metric problem for the second stage.

	Iteration number, k		
	1	2	3
ΔQ (m^3/h)	10.0	9.0	8.1
Z_1^{opt} (MU/h)	36.00	36.00	36.00
Z_2^{opt} (m)	1769.21	1753.23	1751.29
Z_1 (MU/h)	36.00	36.00	36.00
Z_2 (m)	1769.21	1753.23	1751.29
$Lsim$	1766.88	1766.46	1766.46
$\lVert(Q_i^{\,k} - Q_i^{\,k\text{-}1})\rVert$	10000.00	42.43	0.00
Q_1 (extraction well)	50.00	50.00	50.00
Q_2 (extraction well)	50.00	50.00	50.00
Q_3 (extraction well)	50.00	50.00	50.00
Q_4 (extraction well)	50.00	50.00	50.00
Q_5 (extraction well)	50.00	50.00	50.00
Q_6 (extraction well)	50.00	50.00	50.00
Q_7 (extraction well)	0.00	0.00	0.00
Q_8 (extraction well)	0.00	0.00	0.00
Q_9 (extraction well)	0.00	0.00	0.00
Q_{10} (extraction well)	50.00	50.00	50.00
Q_{11}(extraction well)	50.00	50.00	50.00
Q_{12} (extraction well)	50.00	50.00	50.00
Q_{13} (extraction well)	50.00	50.00	50.00
Q_{14} (extraction well)	50.00	50.00	50.00
Q_{15} (extraction well)	50.00	50.00	50.00
Q_{16} (extraction well)	50.00	50.00	50.00
Q_{17} (extraction well)	20.00	50.00	50.00
Q_{18} (extraction well)	0.00	0.00	0.00
Q_{19} (extraction well)	0.00	0.00	0.00
Q_{20} (extraction well)	50.00	20.00	20.00
Q_{21} (injection well)	0.00	0.00	0.00
Q_{22} (injection well)	0.00	0.00	0.00
Q_{23} (injection well)	0.00	0.00	0.00
Q_{24} (injection well)	0.00	0.00	0.00
Q_{25} (injection well)	0.00	0.00	0.00
Q_{26} (injection well)	0.00	0.00	0.00
Q_{27} (injection well)	0.00	0.00	0.00
Q_{28} (injection well)	0.00	0.00	0.00
Q_{29} (injection well)	0.00	0.00	0.00
Q_{30} (injection well)	0.00	0.00	0.00
Q_{31} (injection well)	0.00	0.00	0.00
Q_{32} (injection well)	0.00	0.00	0.00
Q_{33} (injection well)	0.00	0.00	0.00
Q_{34} (injection well)	0.00	0.00	0.00
Q_{35} (injection well)	0.00	0.00	0.00
Q_{36} (injection well)	0.00	0.00	0.00
Q_{37} (injection well)	0.00	0.00	0.00
Q_{38} (injection well)	0.00	0.00	0.00
Q_{39} (injection well)	0.00	0.00	0.00
Q_{40} (injection well)	0.00	0.00	0.00

Table 7.9: The optimal values and the $Lsim$ at all different iterations (L_1-metric).

Iteration	ΔQ (m^3/h)	Z_1^{opt} (MU/h)	Z_2^{opt} (m)	Z_1 (MU/h)	Z_2 (m)	$Lsim$ (m)	$\|(Q_i^k - Q_i^{k-1})\|$
1	10.0	36.00	1769.21	36.00	1769.21	1766.88	10000.00
2	9.0	36.00	1753.23	36.00	1753.23	1766.46	42.43
3	8.1	36.00	1751.29	36.00	1751.29	1766.46	0.00

Table 7.10: The optimal solution at the last iteration (L_1-metric)

Well Number	Extraction wells			Well Number	Injection wells		
	Row	Column	Q_{opt} (m^3/h)		Row	Column	Q_{opt} (m^3/h)
1	3	9	50.00	21	2	16	0.000
2	5	16	50.00	22	4	19	0.000
3	7	19	50.00	23	6	24	0.000
4	9	23	50.00	24	8	30	0.000
5	11	28	50.00	25	10	33	0.000
6	13	38	50.00	26	12	41	0.000
7	15	43	0.00	27	14	45	0.000
8	17	63	0.00	28	16	58	0.000
9	19	70	0.00	29	18	74	0.000
10	21	73	50.00	30	20	78	0.000
11	2	10	50.00	31	3	20	0.000
12	4	13	50.00	32	5	27	0.000
13	6	18	50.00	33	7	30	0.000
14	8	16	50.00	34	9	34	0.000
15	10	27	50.00	35	11	39	0.000
16	12	35	50.00	36	13	49	0.000
17	14	39	50.00	37	15	54	0.000
18	16	53	0.00	38	17	74	0.000
19	18	68	0.00	39	19	81	0.000
20	20	72	20.00	40	21	84	0.000

- ***The results of the multi-objective optimal solution by the weighting method ($w = \infty$)***

When $w = \infty$, the weighted multi-objective problem is equivalent to the single Z_2 objective problem. This is because the first term in the right-hand side of the equation:

$$Z = w_1 Z_1 + w_2 Z_2$$

is dropped (it is due to $w_1 = 0$ so that the ratio $w_2/w_1 = w = 1/0 \rightarrow \infty$).

Therefore, the optimal solution of the weighted multi-problem is exactly the same as the solution of the single Z_2 objective problem.

1. The optimal solution $x = \boldsymbol{Q}_{opt}$, in the two iterations are shown in Table 7.11. The ΔQ is initially equal to 10 m^3/h and gradually decreased by multiplying ΔQ with 0.9. The problem converges at the second iteration.

2. The optimal values of Z_1 and Z_2 and the SW intrusion increment, *Lsim*, which are shown in Table 7.12, are based on the optimal solution of the $w = \infty$ weighted problem at different iterations. At the second iteration, the optimal value Z_2 is 2017.41 m, which indicates that the interface toe is moving land-ward. The simulated saltwater intrusion increment, *Lsim*, is also a positive value (2014.66 m) which is roughly the same as the optimal value Z_2 and $Z_2 > Lsim$. In this second stage, the optimal values of the weighted problem are in the non-inferior set, which is also a point with co-ordinates as depicted in Figure 7.17.

3. The optimal solution $x = \boldsymbol{Q}_{opt}$, which is shown in Table 7.13, is obtained at the second iteration ($k = 2$) with the 15 active extraction wells. Most of these wells have the maximum well capacity (50 m^3/h) except for the well number 12 with the smallest rate of 20m^3/h allocated in the south. Most of those wells are located in the south and middle parts. The locations of active wells are the same as in the L_1-metric problem. The summation of these pumping rates is equal to the water demand ($W_D = 720$ m^3/h). These values of $\boldsymbol{Q}_{opt}$ can be referred to in Figure 7.18 which is the output file of SHARP for mapping the extent of SW intrusion based on the last-iteration optimal solution vector.

4. The SW intrusion interface simulated by SHARP computer code

By comparing Figure 7.13, which shows the initial interfaces at the beginning of the second stage, with Figure 7.18, it becomes obvious that the saltwater toe approaches toward the well locations at the end of the second stage.

Discussion

If the result map (Figure 7.18) of the end of the second stage is compared with Figure 7.4, which shows the initial interface at the beginning of the first stage, it can be seen that their interfaces are nearly the same as each other. However, the interface at the end of the second stage in the $w = \infty$ weighted problem is farther inland than in the L_1-metric problem. In this problem the simulated freshwater zone is extended by an average distance of 516.71 m after management of one year. This value is calculated by the difference between the simulated saltwater intrusion increments of the first stage and the second stage (i.e. –2531.37 m and 2014.66 m, respectively). If calculated by the different optimal values at the end of the two stages (i.e. –2644.69 m and 2017.41 m, respectively), the average distance is about 627.28 m.

Similar to the L_1-metric problem at the second stage, the optimal values of the weighted problem are coordinated in the non-inferior set. This is also a point, i.e. $(Z_1, Z_2) = (36, 2017.41)$, which coincides with the utopia point.

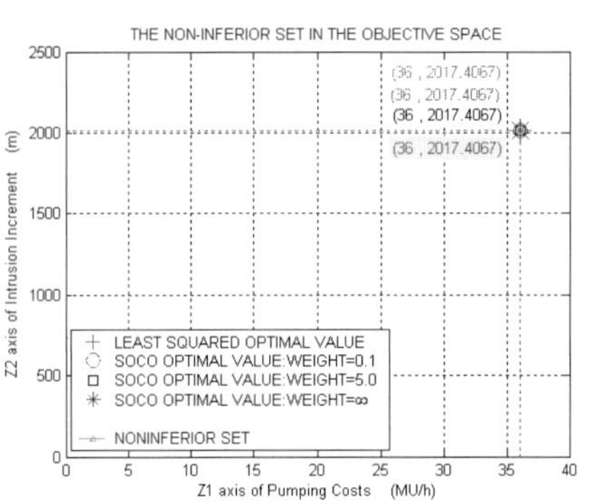

Figure 7.17: The second iteration.

Table 7.11: The results of the weighted problem for the second stage ($w = \infty$).

	Iteration number, k	
	1	2
ΔQ (m^3/h)	10.00	9.00
Z_1^{opt} (MU/h)	36.00	36.00
Z_2^{opt} (m)	2016.34	2017.41
Z_1 (MU/h)	36.00	36.00
Z_2 (m)	2017.41	2017.41
$Lsim$	2014.66	2014.66
$\|(Q_i^k - Q_i^{k-1})\|$	10000.00	0.00
Q_1 (extraction well)	50.00	50.00
Q_2 (extraction well)	50.00	50.00
Q_3 (extraction well)	50.00	50.00
Q_4 (extraction well)	50.00	50.00
Q_5 (extraction well)	50.00	50.00
Q_6 (extraction well)	50.00	50.00
Q_7 (extraction well)	0.00	0.00
Q_8 (extraction well)	0.00	0.00
Q_9 (extraction well)	0.00	0.00
Q_{10} (extraction well)	50.00	50.00
Q_{11}(extraction well)	50.00	50.00
Q_{12} (extraction well)	20.00	20.00
Q_{13} (extraction well)	50.00	50.00
Q_{14} (extraction well)	50.00	50.00
Q_{15} (extraction well)	50.00	50.00
Q_{16} (extraction well)	50.00	50.00
Q_{17} (extraction well)	50.00	50.00
Q_{18} (extraction well)	0.00	0.00
Q_{19} (extraction well)	0.00	0.00
Q_{20} (extraction well)	50.00	50.00
Q_{21} (injection well)	0.00	0.00
Q_{22} (injection well)	0.00	0.00
Q_{23} (injection well)	0.00	0.00
Q_{24} (injection well)	0.00	0.00
Q_{25} (injection well)	0.00	0.00
Q_{26} (injection well)	0.00	0.00
Q_{27} (injection well)	0.00	0.00
Q_{28} (injection well)	0.00	0.00
Q_{29} (injection well)	0.00	0.00
Q_{30} (injection well)	0.00	0.00
Q_{31} (injection well)	0.00	0.00
Q_{32} (injection well)	0.00	0.00
Q_{33} (injection well)	0.00	0.00
Q_{34} (injection well)	0.00	0.00
Q_{35} (injection well)	0.00	0.00
Q_{36} (injection well)	0.00	0.00
Q_{37} (injection well)	0.00	0.00
Q_{38} (injection well)	0.00	0.00
Q_{39} (injection well)	0.00	0.00
Q_{40} (injection well)	0.00	0.00

Table 7.12: The optimal values and the *Lsim* at all different iterations ($w = \infty$).

Iteration	ΔQ (m^3/h)	Z_1^{opt} (MU/h)	Z_2^{opt} (m)	Z_1 (MU/h)	Z_2 (m)	*Lsim* (m)	$\|(Q_i^k - Q_i^{k-1})\|$
1	10.0	36.00	2016.34	36.00	2017.41	2014.66	10000.00
2	9.0	36.00	2017.41	36.00	2017.41	2014.66	0.00

Table 7.13: The optimal solution at the last iteration ($w = \infty$)

Well	Extraction wells			Well	Injection wells		
Number	Row	Column	Q_{opt} (m^3/h)	Number	Row	Column	Q_{opt} (m^3/h)
1	3	9	50.00	21	2	16	0.000
2	5	16	50.00	22	4	19	0.000
3	7	19	50.00	23	6	24	0.000
4	9	23	50.00	24	8	30	0.000
5	11	28	50.00	25	10	33	0.000
6	13	38	50.00	26	12	41	0.000
7	15	43	0.00	27	14	45	0.000
8	17	63	0.00	28	16	58	0.000
9	19	70	0.00	29	18	74	0.000
10	21	73	50.00	30	20	78	0.000
11	2	10	50.00	31	3	20	0.000
12	4	13	20.00	32	5	27	0.000
13	6	18	50.00	33	7	30	0.000
14	8	16	50.00	34	9	34	0.000
15	10	27	50.00	35	11	39	0.000
16	12	35	50.00	36	13	49	0.000
17	14	39	50.00	37	15	54	0.000
18	16	53	0.00	38	17	74	0.000
19	18	68	0.00	39	19	81	0.000
20	20	72	50.00	40	21	84	0.000
20	20	72	50.00	40	21	84	0.000

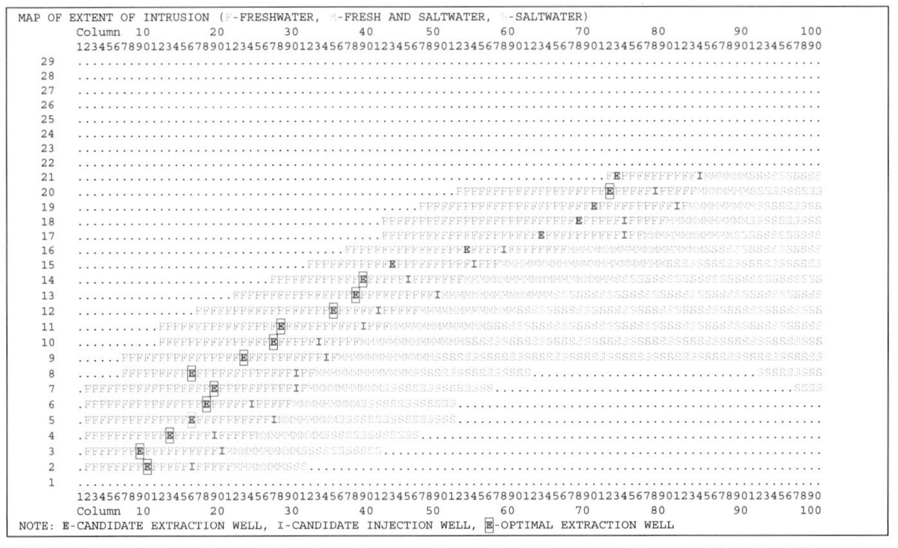

Figure 7.18: Plan view of the interface at the end of the second stage ($k = 5$ m/d).

7.4.3 Conclusions for the deterministic problems

For the first stage, the optimal value Z_2 of the weighted problem with $w = 1.3$ is (in absolute value) bigger than that of the L_2-metric problem and even the L_1-metric value (see Figure 7.6). This shows that the weighting method can be used for instances where decision-makers want to support the Z_2 objective much more in the multi-objective problem.

Also in the first stage, the optimal value Z_2 is smaller than the simulated saltwater intrusion increment for the weighted problem and, inversely, the optimal value Z_2 is greater than the simulated saltwater intrusion increment for the L_2-metric problem. Therefore, the optimal value Z_2 of the L_2-metric problem is overestimated and the weighted problem is underestimated in comparison with their corresponding simulated results. If the simulated results are assumed to be realistic, the criterion such as $Z_2 \geq Lsim$ should be satisfied in any cases. This criterion guarantees that once the project is implemented, the resulting saltwater intrusion length is always smaller than the optimal saltwater intrusion length. Thus, it can be said that, in the first stage, the result of the L_2-metric problem is more satisfactory than the result of the weighted problem for the implementation, in terms of safety levels.

In the second stage, inversely, the results of the $w = \infty$ weighted problem satisfy the condition $Z_2 \geq Lsim$, whereas the results of the L_1-metric problem do not. However, the differences between the optimal values Z_2 and the simulated values $Lsim$ in two problems are all small (< 1%).

In the second stage, the non-inferior set is a point that coincides with the point of the optimal values of the multi-objective problems formulated by either the L_1-metric or the $w = \infty$ weighting methods.

For the deterministic case, the saltwater intrusion management scheme proposed is to apply the weighted problem for the first stage and either the L_1-metric problem or the weighted problem ($w = \infty$) for the second stage. After one year of applying the management scheme, the results of the model show that the freshwater zone will gain some distance for both the weighted and L_1-metric problems. If using the L_1-metric method for the second stage, the freshwater zone will gain an average distance of 764.91 m (or 893.40 m by the optimization computation). It is greater than the average distance of 516.71 m (or 627.28 m by the optimization computation) attained by the weighted problem with $w = \infty$ for the second stage.

For both problems the interface toe (saltwater zone) does not pass any well locations in any stages.

7.5 Results of the robust multi-objective problem under the hydraulic conductivity uncertainty

7.5.1 The random values of the hydraulic conductivity

In this problem, the random values of the hydraulic conductivity are generated by the sequential Gausian simulation approach with the kriging variance (*cc*) of 1.0. The maximum value of the hydraulic conductivity is 35 m/d and the minimum one is 4.5 m/d. The total number of realizations generated for the problem is 20. These random fields are simulated on a grid of 29×66 nodes, with the total grid measuring 36250 m × 82500 m, so the cell size is 1250 m × 1250 m. The random fields will then be transformed into the modeled area with a particular shape and different-size cells. The two-dimensional plot of one of these particular random fields is drawn in Figure 7.19.

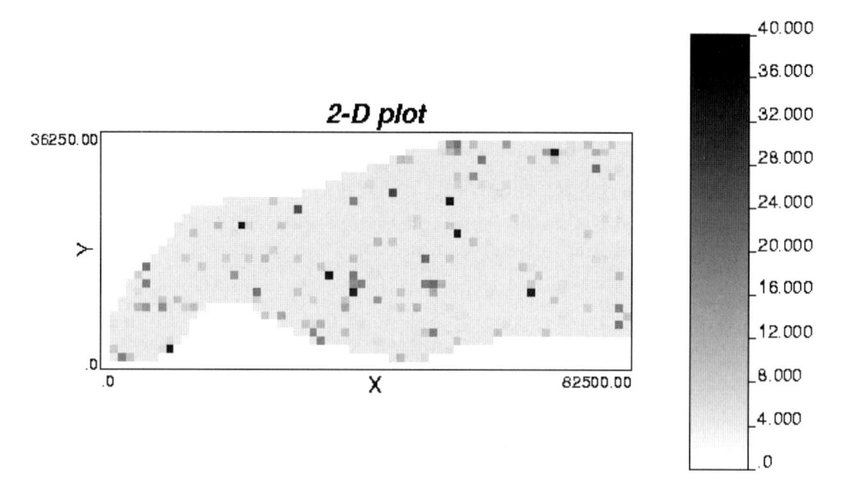

Figure 7.19: The hydraulic conductivity distribution of the 20th realization.

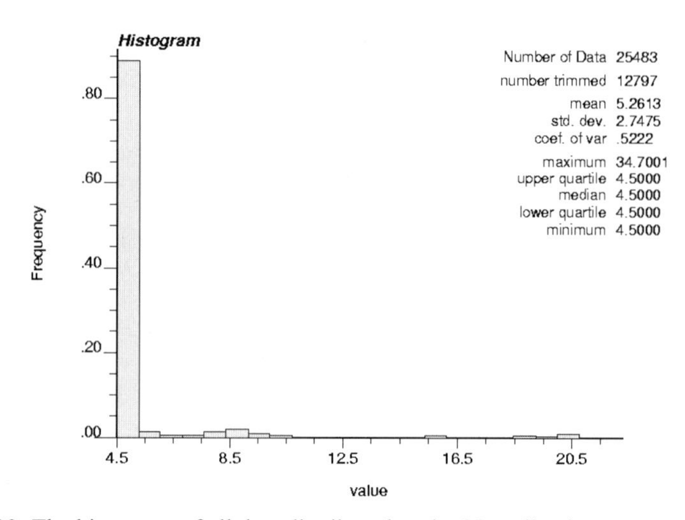

Figure 7.20: The histogram of all data distributed to the 20 realizations.

At the non-pumping condition, the interface of the modelled area is simulated based on each of the 20 realizations of hydraulic conductivity. The particular interface position of the twentieth realization is shown on the map in Figure 7.21. This result map, that consists of only 100 columns of size 250m in the x direction, is extracted from the whole modelled area map of the output file.

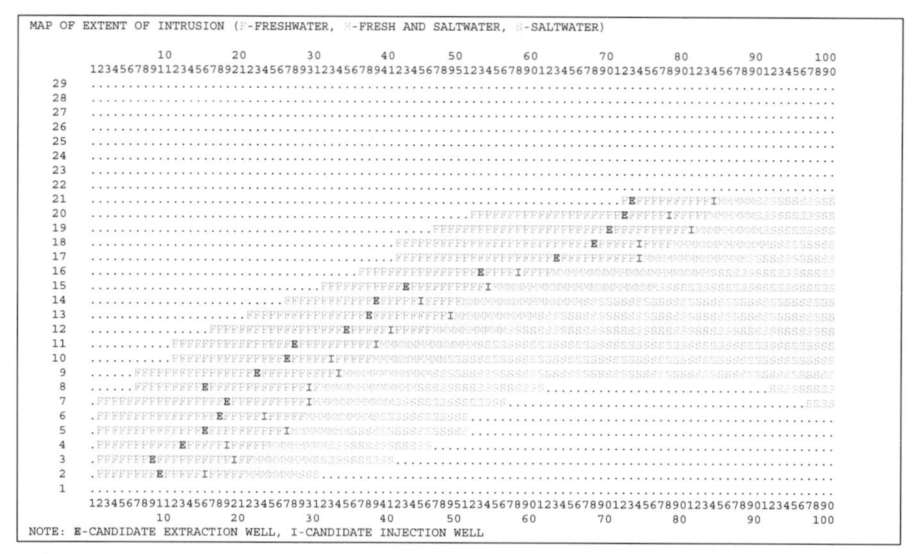

Figure 7.21: The interface at the non-pumping condition for the wet season (for the 20th realization).

7.5.2 The saltwater intrusion management in the wet season (the first stage)

- ***The results of the multi-objective optimal solution by the L_2-metric method***

1. The optimal solution, $x = \boldsymbol{Q}_{opt}$, converges at the second iteration. The optimal solutions of two iterations are all distributed to the same 18 active wells, i.e. eight extraction wells and 10 injection wells as shown in Table 7.14. In that table the simulated saltwater intrusion increments are computed as in a range of (–2076.87, –1794.57).

2. The optimal values of (Z_1, Z_2) are computed as (918, –1777.26) at the second iteration. The minimum and maximum values of the simulated saltwater intrusion increments are –2076.87 m and –1794.57 m, respectively. This maximum value is nearly the same as the optimal value Z_2. This means the optimal value Z_2 is outside and to the right of the uncertainty range of the simulated saltwater intrusion increments. This shows the overestimation of the optimal value Z_2 to the realizations of the simulated saltwater intrusion increments. This overestimation ensures the project, if implemented, will satisfy the criterion such as $Z_2 \geq Lsim_{\xi}$. This shows the high level of safety or say, robustness, in the uncertainty case of the L_2-metric problem.

The mean value of this range, $Lsim\_avg$, is equal to –1973.33 m as shown in Table 7.15. The trade-off values of (Z_1, Z_2) in the non-inferior set at the first and last iterations are depicted in Figures 7.22 and 7.23.

3. The optimal solution $x = \boldsymbol{Q}_{opt}$, which is shown in Table 7.16, consists of 18 active wells i.e. eight extraction wells and 10 injection wells. In the wet season, if comparing this solution with the L_2-metric solution in the deterministic problem, it is obvious that the two optimal solutions are the same in terms of the active well number, rates and locations. However the optimal value Z_2 (−1777.26) of this problem is bigger (smaller if in absolute values) than the one in the deterministic problem. These can prove for the L_2-metric problems that the uncertainty of the hydraulic conductivity influences the optimal values and that the variation of the boundary flow affects the optimal solution.

4. The output file of SHARP for mapping the extent of SW intrusion based on the optimal solution is shown in Figure 7.24.

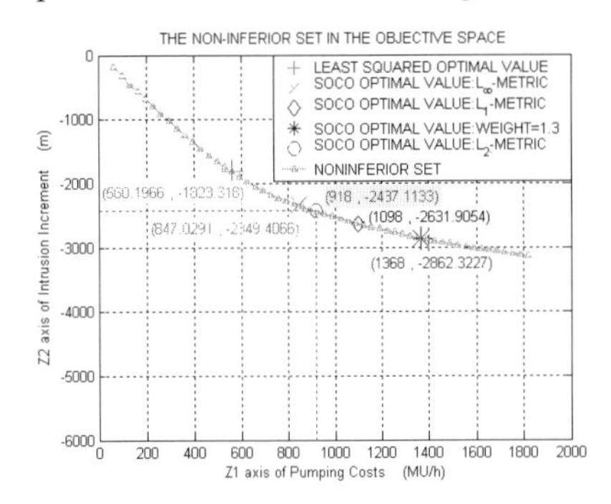

Figure 7.22: The first iteration.

Figure 7.23: The second iteration.

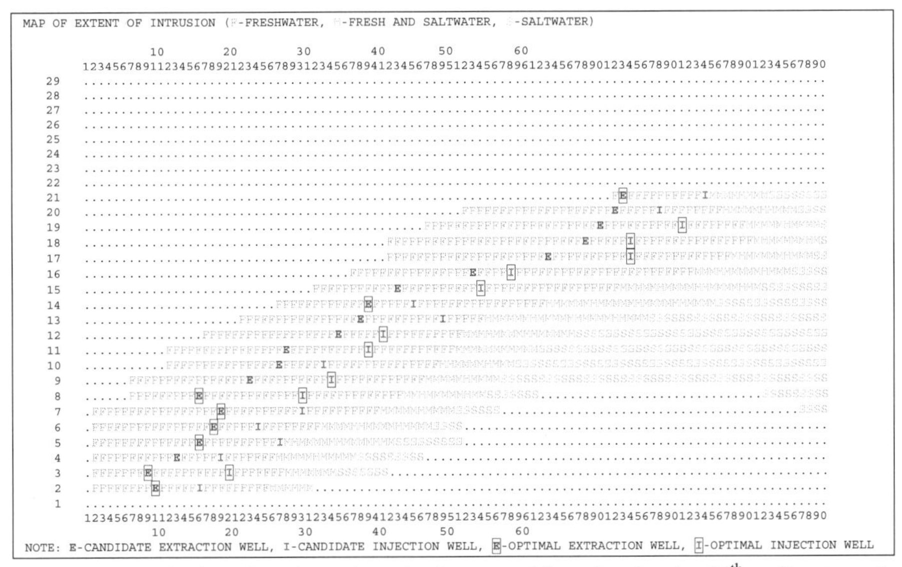

Figure 7.24: The interface tip and toe in the second iteration for the 20th realization of the hydraulic conductivity values.

Table 7.14: The results of the L_2-metric problem in the first stage under uncertainty case.

	Iteration number, k			Iteration number, k	
	1	2		1	2
ΔQ (m^3/h)	10.000	9.800			
Z_1 (MU/h)	918.000	918.000			
Z_2 (m)	-2437.113	-1777.264			
$Lsim_1$	-1915.518	-1915.518			
$Lsim_2$	-1834.379	-1834.379			
$Lsim_3$	-2017.646	-2017.646			
$Lsim_4$	-1794.585	-1794.573			
$Lsim_5$	-2027.997	-2027.997			
$Lsim_6$	-1968.548	-1968.548			
$Lsim_7$	-2054.123	-2054.123			
$Lsim_8$	-1913.936	-1913.928			
$Lsim_9$	-2076.838	-2076.870			
$Lsim_{10}$	-1999.042	-1999.051			
$Lsim_{11}$	-1944.663	-1944.603			
$Lsim_{12}$	-2052.115	-2052.151			
$Lsim_{13}$	-1914.039	-1914.039			
$Lsim_{14}$	-2043.481	-2043.481			
$Lsim_{15}$	-1941.812	-1941.812			
$Lsim_{16}$	-2014.530	-2014.530			
$Lsim_{17}$	-1937.220	-1937.223			
$Lsim_{18}$	-1950.290	-1950.358			
$Lsim_{19}$	-2024.504	-2024.468			
$Lsim_{20}$	-2041.234	-2041.235			
$\|\|(Q_i^{\,k} - Q_i^{\,k-1})\|\|$	10000.000	0.000			
Q_1 (extraction well)	50.000	50.000	Q_{21} (injection well)	0.000	0.000
Q_2 (extraction well)	50.000	50.000	Q_{22} (injection well)	0.000	0.000
Q_3 (extraction well)	50.000	50.000	Q_{23} (injection well)	0.000	0.000
Q_4 (extraction well)	0.000	0.000	Q_{24} (injection well)	900.000	900.000
Q_5 (extraction well)	0.000	0.000	Q_{25} (injection well)	0.000	0.000
Q_6 (extraction well)	0.000	0.000	Q_{26} (injection well)	900.000	900.000
Q_7 (extraction well)	0.000	0.000	Q_{27} (injection well)	0.000	0.000
Q_8 (extraction well)	0.000	0.000	Q_{28} (injection well)	900.000	900.000
Q_9 (extraction well)	0.000	0.000	Q_{29} (injection well)	900.000	900.000
Q_{10} (extraction well)	10.000	10.000	Q_{30} (injection well)	0.000	0.000
Q_{11}(extraction well)	50.000	50.000	Q_{31} (injection well)	900.000	900.000
Q_{12} (extraction well)	0.000	0.000	Q_{32} (injection well)	0.000	0.000
Q_{13} (extraction well)	50.000	50.000	Q_{33} (injection well)	0.000	0.000
Q_{14} (extraction well)	50.000	50.000	Q_{34} (injection well)	900.000	900.000
Q_{15} (extraction well)	0.000	0.000	Q_{35} (injection well)	900.000	900.000
Q_{16} (extraction well)	0.000	0.000	Q_{36} (injection well)	0.000	0.000
Q_{17} (extraction well)	50.000	50.000	Q_{37} (injection well)	900.000	900.000
Q_{18} (extraction well)	0.000	0.000	Q_{38} (injection well)	900.000	900.000
Q_{19} (extraction well)	0.000	0.000	Q_{39} (injection well)	900.000	900.000
Q_{20} (extraction well)	0.000	0.000	Q_{40} (injection well)	0.000	0.000

Table 7.15: The optimal values and the *Lsim-avg* at all different iterations (L_2-metric).

Iteration	ΔQ (m^3/h)	Z_1 (MU/h)	Z_2 (m)	*Lsim_avg* (m)	$\|\|(Q_i^{\ k} - Q_i^{\ k-1})\|\|$
1	10.00	918.000	–2437.113	–1973.325	10000.000
2	9.80	918.000	–1777.264	–1973.327	0.000

Table 7.16: The optimal solution of the uncertainty problem with L_2-metric method.

Well Number	Extraction wells Row	Column	Q_{opt} (m^3/h)	Well Number	Injection wells Row	Column	Q_{opt} (m^3/h)
1	3	9	50.000	21	2	16	0.000
2	5	16	50.000	22	4	19	0.000
3	7	19	50.000	23	6	24	0.000
4	9	23	0.000	24	8	30	900.000
5	11	28	0.000	25	10	33	0.000
6	13	38	0.000	26	12	41	900.000
7	15	43	0.000	27	14	45	0.000
8	17	63	0.000	28	16	58	900.000
9	19	70	0.000	29	18	74	900.000
10	21	73	10.000	30	20	78	0.000
11	2	10	50.000	31	3	20	900.000
12	4	13	0.000	32	5	27	0.000
13	6	18	50.000	33	7	30	0.000
14	8	16	50.000	34	9	34	900.000
15	10	27	0.000	35	11	39	900.000
16	12	35	0.000	36	13	49	0.000
17	14	39	50.000	37	15	54	900.000
18	16	53	0.000	38	17	74	900.000
19	18	68	0.000	39	19	81	900.000
20	20	72	0.000	40	21	84	0.000

- ***The results of the multi-objective optimal solution by the weighted problem (w = 1.3)***

1. The optimal solution $x = \boldsymbol{Q}_{opt}$, converges after two iterations. The optimal solution is allocated to 23 active wells i.e. eight extraction wells and 15 injection wells as shown in Table 7.17.

2. The optimal values of (Z_1, Z_2) are computed as (1368, –2630.80). These optimal values can be seen as a trade-off (in favour of the Z_2 objective) with the optimal values, (918, –1777.264), in the L_2-metric method. The twenty simulated SW intrusion increments, $Lsim_\xi$, are in the range of (–2793.43, –2385.05) as shown in Table 7.17, and the mean value, *Lsim_avg* = –2618.62m, is shown in Table 7.18. If comparing the optimal value Z_2 with the simulated SW intrusion increments, then Z_2 (–2630.80 m) is more or less the same as the mean value *Lsim_avg* of these increments. This means the optimal value Z_2 is in the middle of the uncertainty range of the simulated saltwater intrusion increments. This shows the optimal value Z_2 is not always greater than all the realizations of $Lsim_\xi$ values and the criterion, $Z_2 \geq Lsim_\xi$, is not always satisfied. It can be said that the robustness in the uncertainty case of the weighted problem (with w = 1.3) is less than in the L_2-metric problem.

The trade-off values of (Z_1, Z_2) in the non-inferior set at the first and last iteration are depicted in Figures 7.25 and 7.26. The non-inferior sets drawn in those figures are created by the constraint method with the response matrices [ΔH_i / ΔQ_j] and [ΔL / ΔQ_j]. These are based on the optimal solution from the w = 1.3 weighted problem. The optimal values with the weights w = 0.1 and w = 5.0 in the non-inferior set are just for showing the full range of choice in the non-inferior set for the weighted problem.

3. The optimal solution $x = \boldsymbol{Q}_{opt}$, which is shown in Table 7.19, is obtained at the second iteration (k = 2). If compared with the L_2-metric optimal solution in the uncertainty case, the extraction wells are the same in both cases, but the number of injection wells in the weighted problem is five wells more than in the L_2-metric problem. This also means the weighted problem with w = 1.3 is more in favour of the Z_2 objective than the L_2-metric problem.

If comparing the optimal solution of this weighted problem with the one in the deterministic case, the extraction wells are also the same in both cases but the number of injection wells in the uncertainty case is one well more than in the deterministic case. Although, the optimal value Z_2 (–2630.80 m) in the uncertainty case is slightly larger (smaller if its absolute value) than the Z_2 (–2644.69 m) in the deterministic case. This shows that in the uncertainty case, more injection wells need to be performed, hence, the optimal value Z_1 will increase if the same objective value Z_2 is to be achieved as in the deterministic case.

The reader can refer these values of $\boldsymbol{Q}_{opt}$ to Figure 7.27, which is the output file of SHARP for mapping the extent of SW intrusion.

4. In Figure 7.27, the distribution of wells is the same as in the weighted problem of the deterministic case (Figure 7.10) except for the 30th injection well (row 20, column 78) that is added in the uncertainty case. This additional injection well will result in a certain extension of the freshwater zone in the northern part. In contrast, in the rest of the area, the interfaces of both cases are more or less the same. Of course, if compared with the L_2-metric problem (Figure 7.24) in this uncertainty case, the freshwater zone resulted from the weighted problem is extended further seaward for the whole area.

Discussion

If compared with the L_2-metric problem results, there are 9 out of 14 injection wells of the weighted problem which are more predominantly distributed in the middle area (i.e. in the ratio of 9 to 14 if compared with the ratio of 6 to 10). The other three injection wells are allocated to the north, as in the L_2-metric problem, whereas one more injection well is added to the southernmost area for the weighted problem. The four additional injection wells with such a distribution drive back the interfaces seaward, which results in the expansion of the freshwater zone in the southernmost area.

With the greater values of w the weighted problem will give optimal values which are more advantageous to the objective Z_2 and consequently, the second stage of management problem will be more feasible in the sense of the saltwater intrusion.

The L_2-metric problem is more robust than the weighted problem. It is more necessary, especially, for the second stage (in the dry season) where calamities such as shortage of fresh surface water will result in more possibilities for the interface to reach the capture zone.

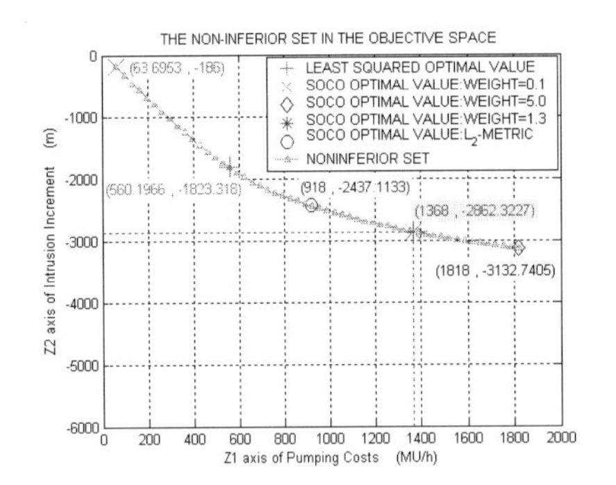

Figure 7.25: The first iteration.

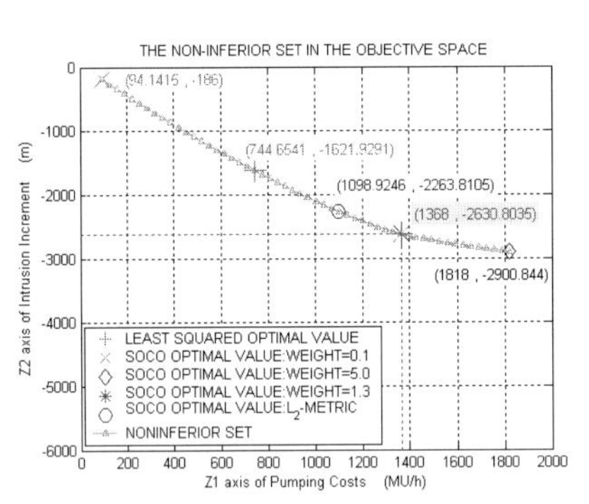

Figure 7.26: The second iteration.

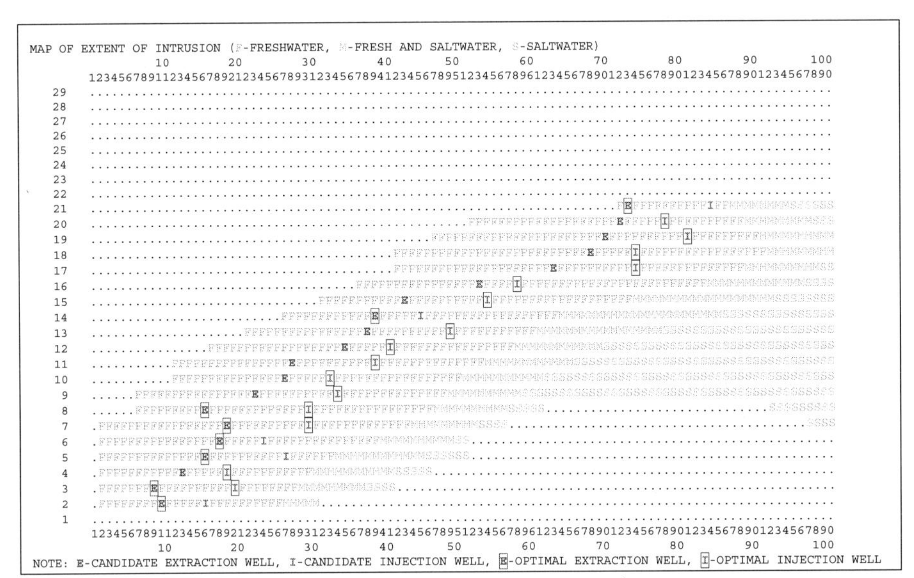

Figure 7.27: The interface in the second iteration for the 20th realization.

Table 7.17: The results of weighted problem (w = 1.3) in the first stage under uncertainty case.

	Iteration number, k			Iteration number, k	
	1	2		1	2
ΔQ (m^3/h)	10.000	9.800			
Z_1 (MU/h)	1368.000	1368.000			
Z_2 (m)	-2862.323	-2630.804			
$Lsim_1$	-2633.285	-2633.285			
$Lsim_2$	-2602.345	-2602.343			
$Lsim_3$	-2561.944	-2561.944			
$Lsim_4$	-2426.416	-2426.424			
$Lsim_5$	-2699.427	-2699.427			
$Lsim_6$	-2657.839	-2657.827			
$Lsim_7$	-2793.434	-2793.434			
$Lsim_8$	-2611.534	-2611.502			
$Lsim_9$	-2694.300	-2694.295			
$Lsim_{10}$	-2678.453	-2678.453			
$Lsim_{11}$	-2633.261	-2633.263			
$Lsim_{12}$	-2719.111	-2719.113			
$Lsim_{13}$	-2577.105	-2577.103			
$Lsim_{14}$	-2567.773	-2567.825			
$Lsim_{15}$	-2579.156	-2579.185			
$Lsim_{16}$	-2591.092	-2591.092			
$Lsim_{17}$	-2546.440	-2546.495			
$Lsim_{18}$	-2385.214	-2385.054			
$Lsim_{19}$	-2699.155	-2699.142			
$Lsim_{20}$	-2715.149	-2715.148			
$\Vert(Q_i^{\,k} - Q_i^{\,k-1})\Vert$	10000.000	0.000			
Q_1 (extraction well)	50.000	50.000	Q_{21} (injection well)	0.000	0.000
Q_2 (extraction well)	50.000	50.000	Q_{22} (injection well)	900.000	900.000
Q_3 (extraction well)	50.000	50.000	Q_{23} (injection well)	0.000	0.000
Q_4 (extraction well)	0.000	0.000	Q_{24} (injection well)	900.000	900.000
Q_5 (extraction well)	0.000	0.000	Q_{25} (injection well)	900.000	900.000
Q_6 (extraction well)	0.000	0.000	Q_{26} (injection well)	900.000	900.000
Q_7 (extraction well)	0.000	0.000	Q_{27} (injection well)	0.000	0.000
Q_8 (extraction well)	0.000	0.000	Q_{28} (injection well)	900.000	900.000
Q_9 (extraction well)	0.000	0.000	Q_{29} (injection well)	900.000	900.000
Q_{10} (extraction well)	10.000	10.000	Q_{30} (injection well)	900.000	900.000
Q_{11}(extraction well)	50.000	50.000	Q_{31} (injection well)	900.000	900.000
Q_{12} (extraction well)	0.000	0.000	Q_{32} (injection well)	0.000	0.000
Q_{13} (extraction well)	50.000	50.000	Q_{33} (injection well)	900.000	900.000
Q_{14} (extraction well)	50.000	50.000	Q_{34} (injection well)	900.000	900.000
Q_{15} (extraction well)	0.000	0.000	Q_{35} (injection well)	900.000	900.000
Q_{16} (extraction well)	0.000	0.000	Q_{36} (injection well)	900.000	900.000
Q_{17} (extraction well)	50.000	50.000	Q_{37} (injection well)	900.000	900.000
Q_{18} (extraction well)	0.000	0.000	Q_{38} (injection well)	900.000	900.000
Q_{19} (extraction well)	0.000	0.000	Q_{39} (injection well)	900.000	900.000
Q_{20} (extraction well)	0.000	0.000	Q_{40} (injection well)	0.000	0.000

Table 7.18: The optimal values and the *Lsim-avg* at all different iterations (w = 1.3).

Iteration	ΔQ (m^3/h)	Z_1 (MU/h)	Z_2 (m)	*Lsim_avg* (m)	$\Vert(Q_i^{\,k} - Q_i^{\,k-1})\Vert$
1	10.00	1368.000	-2862.323	-2618.622	10000.000
2	9.80	1368.000	-2630.804	-2618.618	0.000

Table 7.19: The optimal solution at the last iteration (w = 1.3)

Well	Extraction wells			Well	Injection wells		
Number	Row	Column	Q_{opt} (m^3/h)	Number	Row	Column	Q_{opt} (m^3/h)
1	3	9	50.000	21	2	16	0.000
2	5	16	50.000	22	4	19	900.000
3	7	19	50.000	23	6	24	0.000
4	9	23	0.000	24	8	30	900.000
5	11	28	0.000	25	10	33	900.000
6	13	38	0.000	26	12	41	900.000
7	15	43	0.000	27	14	45	0.000
8	17	63	0.000	28	16	58	900.000
9	19	70	0.000	29	18	74	900.000
10	21	73	10.000	30	20	78	900.000
11	2	10	50.000	31	3	20	900.000
12	4	13	0.000	32	5	27	0.000
13	6	18	50.000	33	7	30	900.000
14	8	16	50.000	34	9	34	900.000
15	10	27	0.000	35	11	39	900.000
16	12	35	0.000	36	13	49	900.000
17	14	39	50.000	37	15	54	900.000
18	16	53	0.000	38	17	74	900.000
19	18	68	0.000	39	19	81	900.000
20	20	72	0.000	40	21	84	0.000

7.5.3 The saltwater intrusion management in the dry season (the second stage)

In the second stage's problems for the uncertainty case, the interface at the end of the wet season (the first stage) in the weighted problem (with w = 1.3) is chosen to be the initial interface of the dry season (the second stage). In this stage the optimal solution of the L_1-metric problem cannot be accurately achieved due to a convergence problem. This is because the optimization problem with the two objectives that are not conflicting cannot give a unique solution for all the iterations in this case. The only optimal solution of the L_2-metric problem, however, can be achieved when the pumping costs of different well locations are all the same as before. With this optimal solution, one can say that the achieved optimal values are not the global minima but merely the local minima for this problem.

- ***The results of the multi-objective optimal solution by the L_2-metric method with a constant pumping cost for all well locations***

1. The optimal solution converges after six iterations ($\|(Q_i^{\ k} - Q_i^{\ k-1})\|$ = 5.37) as shown in Table 7.20.

2. The optimal values of Z_1, Z_2, (36.00, 2386.69), at the sixth iteration for this case are not equal to the values of the utopia point, (36.00, 2275.45), and Z_2 is greater than Z_2^{opt}, as expected. These values are shown in Table 7.21.

The optimal value Z_2 is the positive value of 2386.69m, which indicates that the interface toe moves landward. The simulated saltwater intrusion increments of all realizations, $Lsim_\xi$, are also positive values in the range of (1998.75, 2375.36) as shown in Table 7.20. Z_2 is outside and to the right of that range. This means the optimal value Z_2 is greater than all realizations of the simulated saltwater intrusion increments, $Lsim_\xi$, or $Z_2 > Lsim_\xi$. In this second stage the optimal values are the co-ordinates of a point as depicted in Figure 7.28.

3. The optimal solution $x = \boldsymbol{Q}_{opt}$, which is shown in Table 7.22, is evenly distributed to all the candidate extraction wells. This means the active extraction wells are located in all twenty candidate wells and the extraction rates of all these wells are less than the well capacity (50 m^3/h). The summation of these pumping rates is equal to the water demand (W_D = 720 m^3/h) that is set double for the dry season. These values of $\boldsymbol{Q}_{opt}$ can be referred to in Figure 7.29 which is the output file of SHARP for mapping the extent of SW intrusion based on the last-iteration optimal solution vector.

4. The SW intrusion interface simulated by SHARP computer code

If Figure 7.27, which shows the initial interfaces at the beginning of the second stage, is compared with Figure 7.29, it can be seen that the saltwater toe approaches the well locations at the end of the second stage and that there is no well location passed over by the interface.

Discussion

If the result map (Figure 7.29) at the end of the second stage is compared with Figure 7.21, which shows the initial interface at the beginning of the first stage, it can be seen that the interface toes are nearly the same as each other. However, the interface tip at the end of the second stage is closer to the seashore than the interface tip at the beginning of the first stage.

After management for one year, the freshwater zone limited by the interface tip is extended by an average distance of 244 m, if calculated with the optimal value Z_2, and 424 m, if calculated with the simulated value *Lsim_avg*. The first value of distances is calculated by the difference between the optimal values, Z_2, i.e. –2630.80 m and 2386.69 m of the first stage and the second stage, respectively. The second value is calculated by the difference between the simulated saltwater intrusion increments, i.e. –2618.62 m and 2195.05m, similarly.

Since in the second stage all of the injection wells are switched off then the optimal solution is only applicable for the extraction wells and the summation of their rates equals the water demand for this period. Consequently, the objective Z_1 gets a constant optimal value, i.e. Z_1 = 36 MU/h, for the minimal water demand.

Because the optimal values (Z_1, Z_2) are only achieved from the L_2-metric problem and they are worse than the values of the utopia point, ($Z_1 = Z_1^{opt}$ and $Z_2 > Z_2^{opt}$), hence, the problem gets only a local minima.

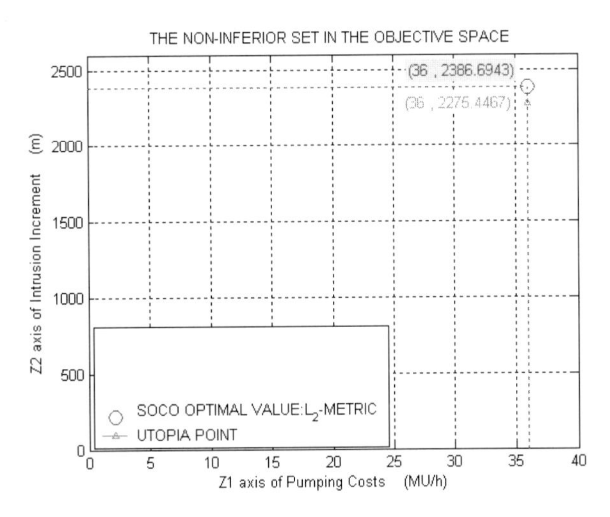

Figure 7.28. The optimal values and the utopia point for the second stage.

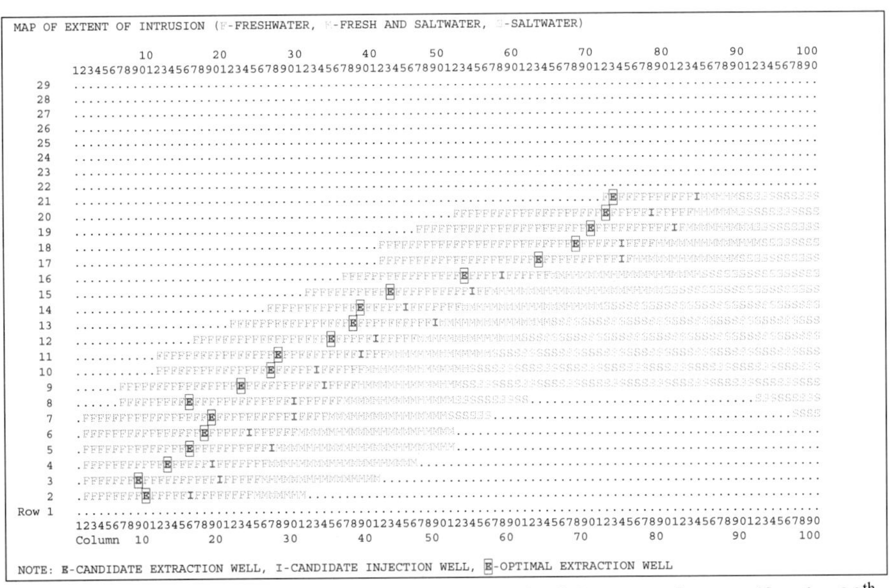

Figure 7.29: Plan view of the interface at the end of the second stage (for the 20th realization).

Table 7.20: The results of the L_2-metric problem for the second stage under uncertainty case.

	Iteration number, k				Iteration number, k		
	1	... 5	6		1	... 5	6
ΔQ (m^3/h)	10.00	6.56	5.90				
Z_1^{opt} (MU/h)	36.00	36.00	36.00				
Z_2^{opt} (m)	2198.63	2235.02	2275.45				
Z_1 (MU/h)	36.00	36.00	36.00				
Z_2 (m)	2236.00	2324.40	2386.69				
$Lsim_1$	2206.83	2206.73	2206.75				
$Lsim_2$	2155.38	2155.98	2155.60				
$Lsim_3$	2106.34	2106.30	2106.66				
$Lsim_4$	2001.75	1998.68	1998.75				
$Lsim_5$	2283.63	2283.71	2286.00				
$Lsim_6$	2205.60	2210.79	2210.86				
$Lsim_7$	2278.58	2278.66	2274.75				
$Lsim_8$	2188.51	2183.13	2183.40				
$Lsim_9$	2221.01	2220.96	2221.14				
$Lsim_{10}$	2291.73	2291.65	2280.70				
$Lsim_{11}$	2370.01	2333.78	2335.01				
$Lsim_{12}$	2259.89	2260.01	2260.36				
$Lsim_{13}$	2139.70	2139.51	2131.58				
$Lsim_{14}$	2165.93	2166.38	2165.41				
$Lsim_{15}$	2375.50	2375.30	2375.36				
$Lsim_{16}$	2112.31	2112.29	2113.00				
$Lsim_{17}$	2083.45	2083.43	2083.71				
$Lsim_{18}$	2002.44	1998.70	1999.06				
$Lsim_{19}$	2248.26	2256.43	2256.95				
$Lsim_{20}$	2255.53	2255.51	2256.00				
$\lVert(Q_i^{\,k} - Q_i^{\,k-1})\rVert$	10000.00	12.16	5.37				
Q_1 (extraction well)	35.90	35.54	35.75	Q_{21} (injection well)	0.00	0.00	0.00
Q_2 (extraction well)	33.80	37.69	36.19	Q_{22} (injection well)	0.00	0.00	0.00
Q_3 (extraction well)	35.23	35.76	35.29	Q_{23} (injection well)	0.00	0.00	0.00
Q_4 (extraction well)	35.86	34.97	36.04	Q_{24} (injection well)	0.00	0.00	0.00
Q_5 (extraction well)	36.55	35.71	32.29	Q_{25} (injection well)	0.00	0.00	0.00
Q_6 (extraction well)	36.48	36.34	35.92	Q_{26} (injection well)	0.00	0.00	0.00
Q_7 (extraction well)	36.60	36.14	36.76	Q_{27} (injection well)	0.00	0.00	0.00
Q_8 (extraction well)	37.07	36.54	37.38	Q_{28} (injection well)	0.00	0.00	0.00
Q_9 (extraction well)	36.30	35.69	35.72	Q_{29} (injection well)	0.00	0.00	0.00
Q_{10} (extraction well)	35.33	35.29	35.29	Q_{30} (injection well)	0.00	0.00	0.00
Q_{11}(extraction well)	36.18	35.61	34.67	Q_{31} (injection well)	0.00	0.00	0.00
Q_{12} (extraction well)	35.66	35.59	36.25	Q_{32} (injection well)	0.00	0.00	0.00
Q_{13} (extraction well)	33.45	32.96	36.08	Q_{33} (injection well)	0.00	0.00	0.00
Q_{14} (extraction well)	36.34	37.52	38.05	Q_{34} (injection well)	0.00	0.00	0.00
Q_{15} (extraction well)	36.64	36.05	35.22	Q_{35} (injection well)	0.00	0.00	0.00
Q_{16} (extraction well)	36.64	36.51	36.54	Q_{36} (injection well)	0.00	0.00	0.00
Q_{17} (extraction well)	36.21	36.10	35.95	Q_{37} (injection well)	0.00	0.00	0.00
Q_{18} (extraction well)	37.06	37.95	38.34	Q_{38} (injection well)	0.00	0.00	0.00
Q_{19} (extraction well)	36.67	36.17	36.45	Q_{39} (injection well)	0.00	0.00	0.00
Q_{20} (extraction well)	36.03	35.86	35.81	Q_{40} (injection well)	0.00	0.00	0.00

Table 7.21: The optimal values and the *Lsim-avg* at all different iterations (L_2-metric).

Iteration	ΔQ (m^3/h)	Z_1^{opt} (MU/h)	Z_2^{opt} (m)	Z_1 (MU/h)	Z_2 (m)	*Lsim_avg* (m)	$\|(Q_i^{\,k} - Q_i^{\,k-1})\|$
1	10.000	36.00	2198.63	36.00	2236.00	2197.62	10000.000
… 5	6.561	36.00	2235.02	36.00	2324.40	2195.90	12.162
6	5.905	36.00	2275.45	36.00	2386.69	2195.05	5.375

Table 7.22: The optimal solution at the sixth iteration (L_2-metric)

Well Number	Extraction wells Row	Column	Q_{opt} (m^3/h)	Well Number	Injection wells Row	Column	Q_{opt} (m^3/h)
1	3	9	35.75	21	2	16	0.000
2	5	16	36.19	22	4	19	0.000
3	7	19	35.29	23	6	24	0.000
4	9	23	36.04	24	8	30	0.000
5	11	28	32.29	25	10	33	0.000
6	13	38	35.92	26	12	41	0.000
7	15	43	36.76	27	14	45	0.000
8	17	63	37.38	28	16	58	0.000
9	19	70	35.72	29	18	74	0.000
10	21	73	35.29	30	20	78	0.000
11	2	10	34.67	31	3	20	0.000
12	4	13	36.25	32	5	27	0.000
13	6	18	36.08	33	7	30	0.000
14	8	16	38.05	34	9	34	0.000
15	10	27	35.22	35	11	39	0.000
16	12	35	36.54	36	13	49	0.000
17	14	39	35.95	37	15	54	0.000
18	16	53	38.34	38	17	74	0.000
19	18	68	36.45	39	19	81	0.000
20	20	72	35.81	40	21	84	0.000

7.5.4. Variation of objective values with kriging variance

The optimization problem given by (7.1)-(7-8) was solved repeatedly by the weighting method (w = 1.3) for only the first managerial stage. But each time the realizations were generated with a different value of kriging variance. This correspondingly results in a different value of mean hydraulic conductivity (for random values of hydraulic conductivity within a range from 4.5m/d to 35m/d). As is shown in Table 7.23, increasing the value of the kriging variance (e.g. cc = 1, 2, 4) results in an increase in the cost and a decrement of SW intrusion of the optimal solution. This is because increasing the kriging variance implies increasing both the heterogeneity level of the distributed hydraulic conductivity (see Figures 7.30, 7.32 and 7.34) and the mean value of the hydraulic conductivity (see Figures 7.31, 7.33 and 7.35). For the first implication an increase in kriging variance is equivalent to an increase in uncertainty, hence optimal solutions corresponding to larger values of kriging variances are likely to be more expensive since they have to guard against higher magnitudes of uncertainty. Moreover, for the second implication, an increase in the mean value of the hydraulic conductivity (e.g. 5.42 m/d, 6.81m/d and 8.94 m/d) does not cause the interface toe moves further seaward (see Figure 3.14). Therefore, a decrement in the SW intrusion length results mainly from an increase in the cost for injection which is caused by the uncertainty as mentioned in the first implication.

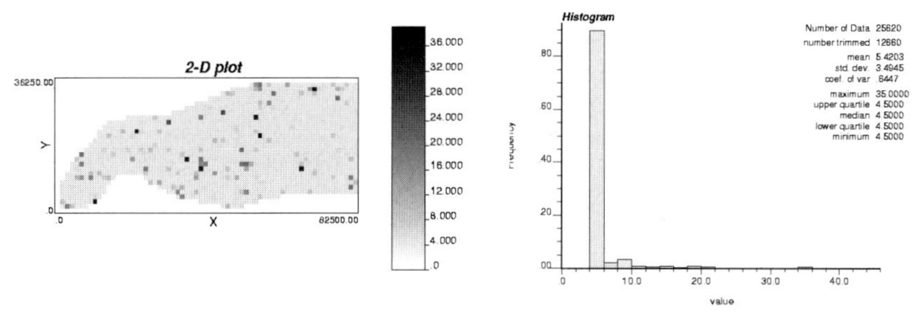

Figure 7.30: One realization of hydraulic conductivity with kriging variance = 1.

Figure 7.31: The histogram of all data. distributed with kriging variance = 1.

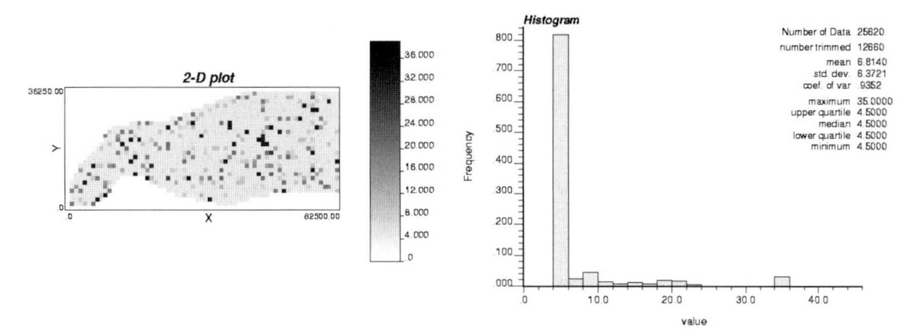

Figure 7.32: One realization of hydraulic conductivity with kriging variance = 2.

Figure 7.33: The histogram of all data. distributed with kriging variance = 2.

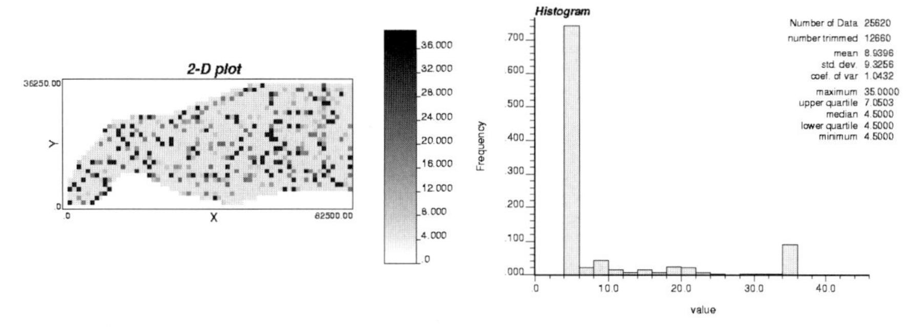

Figure 7.34: One realization of hydraulic conductivity with kriging variance = 4.

Figure 7.35: The histogram of all data. distributed with kriging variance = 4.

Figures 7.36, 7.37 indicate that the relationship between the optimal values of objectives and the kriging variance are not linear. This implies a complication in predicting the optimal values when the hydraulic conductivity is realized with certain kriging variance value.

Figure 7.38 shows how often each well is used for three different kriging variance values. We see that the number of wells that are not used (frequency of 0%) is eleven and there are only four wells which are used one time (frequency of 33.33%). This means that the realized optimal solutions do not support installing of many extra wells but rather reusing of the already existing wells once we implemented one of those optimal strategies.

Table 7.23: Variation of objective values with kriging variance (the hydraulic conductivity is randomly taken within the range from 4.5m/d to 35m/d).

Kriging variance (m^2/d^2)	Mean hydraulic conductivity (m/d)	Cost objective value (*MU*)	SW intrusion decrement (m)
1	5.420	1368.000	2630.804
2	6.814	1638.000	3197.451
4	8.939	1818.000	3384.641

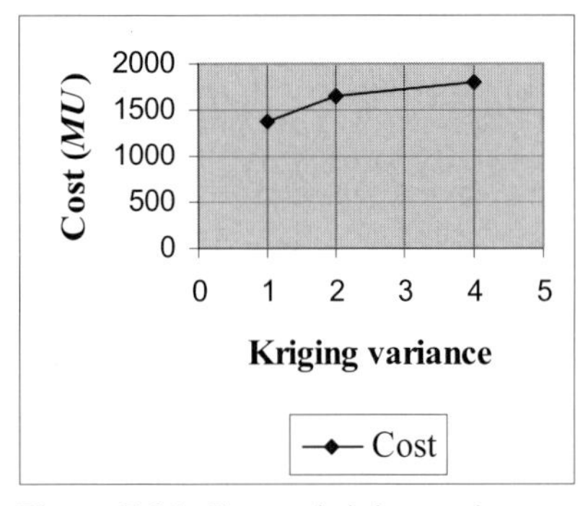

Figure 7.36: Cost vs kriging variance.

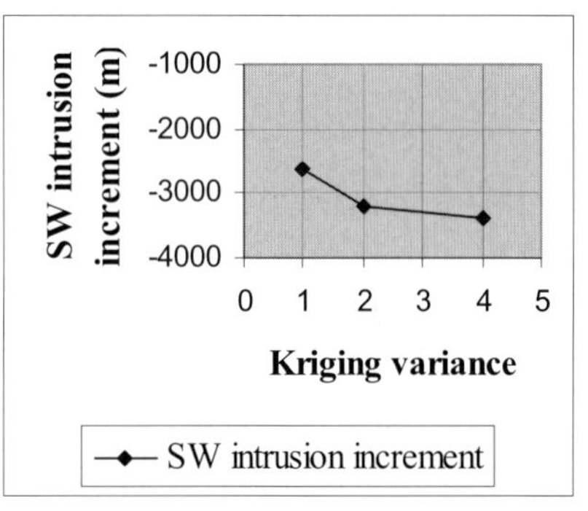

Figure 7.37: SW intrusion increment vs kriging variance.

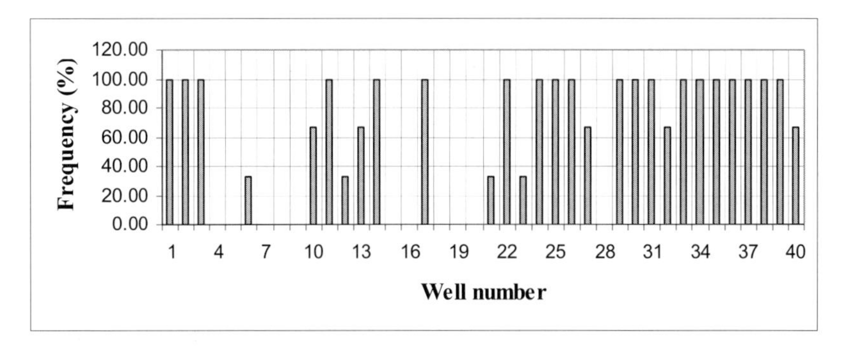

Figure 7.38: Well use frequency for three different kriging variance values.

7.6. Conclusions

In this chapter, how the deterministic and uncertainty problems have been applied to the real-world case model has been examined. Solving the uncertainty problem is rather time-consuming if compared to the corresponding deterministic problem. The time needed for a one-iteration run of the uncertainty case (with 20 realizations) is recorded as at least twelve times greater than for the one of the deterministic case.

By this real-world case application, it can be concluded that:

- The horizontal-averaged saltwater intrusion approach shows its validity in controlling the saltwater intrusion for the asymmetric real-world model by using the multi-objective optimization programs.

- Due to the uncertainty of the aquifer transmissivities, the optimal values of the saltwater intrusion increments in all stages are always greater than the optimal values (Z_2) in the deterministic problems. It means that the interfaces are closer to the capture zone of the optimal wells. This also implies that the risk due to the uncertainty of the aquifer parameter is taken into account in the computation and, hence, the average saltwater intrusion increment predicted is greater than in the deterministic case where there is no risk of the parameter uncertainty. This will help the realization that the saltwater intrusion management of the groundwater aquifer under the uncertainty always has higher costs in order to achieve the same level of saltwater intrusion control.

- Regarding the awareness of the uncertainty risk, the uncertainty problem will trade the time-consumption off against the safety of the optimal solution and, hence, the implementation of the optimized management scheme.

- For the first stage (the wet season), the results of the multi-objective optimization problems show the trade-off among the optimal values (Z_1, Z_2) which are achieved from the two methods, i.e. the weighting method (with given w-weight) and the L_2-metric method. This also shows the flexibility of the weighted problem in adjusting the saltwater intrusion level in the multi-objective management problem by only altering the weight value.

- For the uncertainty problems in the second stage (the dry season), the optimal values that are based on the optimal solution attained from the L_2-metric method are just the local minima. This is because they are only near to the values of the ideal point (utopia point) that could not be achieved when the two objectives proposed are not conflicting in this particular case.

- The small differences between the optimal values, Z_2, and the simulated saltwater intrusion increments for the deterministic problems help to certify the accuracy of the sequential linearization approach that has been used in the deterministic multi-objective optimization programs.

- In the uncertainty problems for both the two stages, the L_2-metric problems show their robustness, which is higher than that of the $w = 1.3$ weighted problems. It is more necessary, especially for the second stage (in the dry season), where calamities such as shortage of fresh surface water will result in more possibilities for the interface to reach the capture zone.

- For the first stage (wet season) we can split up the non-inferior set $\mathbf{N}_o$ in two sets such that for points (Z_1, Z_2) in the first set the MRT is always bigger than for points in the second set. It reflects the fact that application of higher costs for injection is after some point less effective.

- Increasing the value of the kriging variance (e.g. $cc = 1, 2, 4$) results in an increase in the cost and a decrement of SW intrusion of the optimal solution. These are caused by the higher magnitudes of uncertainty when the kriging variance increases.

Chapter 8

Conclusions and Recommendations

8.1 Conclusions

The methodology in this thesis deals with an important issue, which is the uncertainty of the saltwater intrusion management model. The saltwater intrusion management problems of previous studies have encountered some difficulties e.g. reliable data for the simulation model that can be incorporated in the deterministic optimization program. This is because most information sources have been based on the surveying and monitoring of aquifers and salinity of groundwater and, unfortunately, good data is often the weakest point of many studies. Therefore, in this work, the new approach to robust saltwater intrusion management has been successful programming, which enabled the solution of the saltwater intrusion management problems guided by multiple criterions and based on the aquifer parameter uncertainty. The main tasks which have been carried out in this thesis are:

- The simulation model based on the sharp interface approach is applied to a regional scale of the study area. This approach reproduces the regional flow dynamics of the system and response of the interface to applied stresses. Unlike the sharp interface models which simulate flow only in the freshwater region, by incorporating the Ghyben-Herzberg approximation, it is assumed that the saltwater domain adjusts rapidly to applied stresses (see also Ndambuki, 2001). This simulation model reproduces the short-term behavior of a coastal aquifer through the two transient flows in which the influence of saltwater flow is necessarily included. Therefore, it is suitable for the short-term management problems as presented in the real-world case in this work.
- Management of coastal aquifers in this work is guided by two criteria, i.e. location of the interface and the costs of pumping and recharge. In this work the characteristics of the SW intrusion length which are defined by the distance of the location of interface toe from the shoreline have been studied. In this study the horizontally-averaged saltwater intrusion increment response matrix approach is introduced. This approach has been applied and proved its usefulness by enabling the successful control of the saltwater interface under the design stresses in the management problems of this thesis.
- The second-order cone technique is applied to many optimization problems in the saltwater water intrusion management of this work. The optimization problems under varied forms are formulated e.g. the single objective problems, the multi-objective problems with the L_α-metric and weighting methods, and the stochastic optimization problems. The SOCO technique and this work's developed formulas can be used as a guideline for the similar optimization problems.
- Two powerful analysis techniques, i.e. sharp interface simulation and second-order cone optimization have been successfully combined in one program as an add-on for Matlab. The response matrix approach that is applied to both the horizontal-averaged saltwater intrusion increment and head criteria is convenient for use in linking the simulation and optimization programs. It

induces the simplicities in formulating the objectives and constraints in the optimization problems. It produces an engineering design tool, which can be an aid in the formulation of design criteria and constraints, and assists decision-makers in assessing the impacts of design trade-offs.

- The proposed management of saltwater intrusion problems is taken into the computation of the non-linear optimization models due to the non-linear response of the sharp interface with respect to applied stresses. The iterative technique that is introduced to these problems has proved its satisfactory accuracy in linearizing such non-linear problems.

- The second-order cone technique is shown as a useful tool for stochastic programming. Based on this technique, the problems of saltwater intrusion management under uncertainty are successfully formulated and programmed. On the other hand, the deterministic multi-objective saltwater intrusion management is also programmed with the help of this technique to simulate and design the optimal control of flows that have been assumed to be deterministic. The results of these two problems show the differences that depend on the degree of uncertainty (here this is fixed) and the formulation of techniques of the multi-objective problems.

- The results of the stochastic multi-objective problem using the L_2-metric method show the level of robustness that is higher than the one using the weighting method (with $w = 1.3 > 1$). The weighted problems, however, can help decision makers to choose the compromised solution that is in favour of minimizing the saltwater intrusion objective.

- Concerning the non-linear behaviour of the saltwater intrusion response that is involved in both deterministic and uncertainty problems, this requires these programs running iteratively for the convergence. Especially for the stochastic programming, where the program has to repeatedly compute the response matrices of so many realizations in each of the iterations, it needs a considerable CPU time to run the problem until it approaches an accurate solution. However, this is also a trade-off issue when the runtime of the programs has to be reduced with the resulting sacrifice of the accuracy of computer programs.

- The hypothetical problems that show the reasonable results help to check the validity of the computer programs in many cases (e.g. the variants of multi-objective problems solving the deterministic and uncertainty cases).

- The management model is conveniently applied to the real-world case problems in which the two separate management stages for the wet and dry seasons show good results for a promising implementation. The optimal results show the possibility of increasing the number of extraction wells in order to meet the doubled water demand during the dry season without the artificial recharge by injection.

- In the second stage of the management problem, the saltwater intrusion increment due to the transient saltwater flow itself can be seen as a greater threat in comparison with the saltwater intrusion increment due to the extraction purposes. When the injection stops because of calamities, the interface that does not yet achieve the steady-state condition will approach to nearly its initial position after the same time-interval as needed in the injection implementation (see also Stakelbeek, 1999.)

- For the first stage (wet season) we can split up the non-inferior set $\mathbf{N}_o$ in two sets such that for points (Z_1, Z_2) in the first set the MRT is always bigger than for points in the second set. It reflects the fact that application of higher costs for injection is after some point less effective.
- Increasing the value of the kriging variance (e.g. cc = 1, 2, 4) results in an increase in the cost and a decrement of SW intrusion of the optimal solution. These are caused by the higher magnitudes of uncertainty when the kriging variance increases.

8.2 Recommendations

- The new methodology introduced in this work has been developed for the saltwater intrusion management with the multiple criteria. Therefore, it is recommended to use this method for the multi-objective optimization in the saltwater intrusion management problems where the response matrix approach is used for linking the simulation with the optimization.
- Since the method has dealt with the non-linear problems due to the non-linear response of the hydraulic head and the salt/fresh interface with respect to stresses, therefore, it can be applied to either the confined or unconfined aquifers in the groundwater system for controlling the stationary or transient flow.
- The method can also be a useful tool in the multi-objective groundwater quantitative management problems where the saltwater intrusion does not exist and the groundwater system is comprised of confined and unconfined aquifers.
- For practical considerations, it is recommended that the interface of the model be set close to the low concentration isohaline monitored, if the transition zone is negligible in the simulation model.
- The model has been developed for the management problems with the time-invariant case only, hence, it can only be applied separately for each management period of the time-varying case. It is strongly recommended to incorporate the time-varying case into the management model in order to find the optimal solution for over the whole time-varying management problem. This may help to produce low cost solutions (Jae-Heung Yoon and Shoemaker, 1999).
- The run-time can be considerably reduced if reducing the uncertainty. This is recommended to result in the decrease of realization numbers so that the management model can achieve results in less CPU time.
- The criterion of the saltwater intrusion in the model has been based on the horizontally-averaged saltwater intrusion increment to formulate its objective and constraint. Therefore, the model should incorporate the saltwater intrusion control rows into the saltwater intrusion constraints in order to completely control the interface as desired.
- The operational costs that have been used in these multi-objective problems are simplified by neglecting the cost variation due to the different well locations. The variation of costs for the different well locations that can be set in the objective is recommended in order to make the management problems more

realistic. This might also improve the convergence process when solving these problems in the dry season.

- For the second stage, a new objective that is for maximizing the water extraction is recommended to replace the objective for minimizing the SW intrusion increment. The new objective will fully conflict with the operational cost objective. By maximizing the extraction amount the operational cost will increase and this will avoid the alternative optima in the multi-objective problems. This is necessary for a better convergence in the uncertainty problems where the optimal solution cannot be found uniquely for most of the iterations if the operational cost objective remains constant. Preliminary computations have confirmed these expectations.
- For the practical application, the conflicts of the objectives in the multi-objective optimization problems should be carefully checked before applying these to the management problems. This is due to the fact that all the objectives do not always conflict with one another in all managerial stages.

Appendix

Colour figures of the thesis

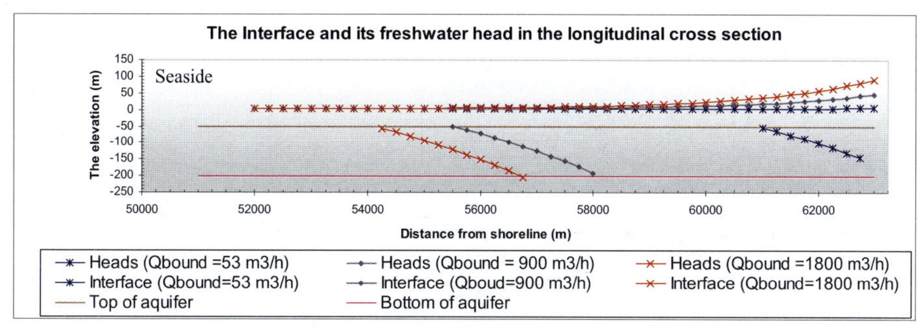

Figure 3.3: The responses of the freshwater head and interface to the boundary flow rates.

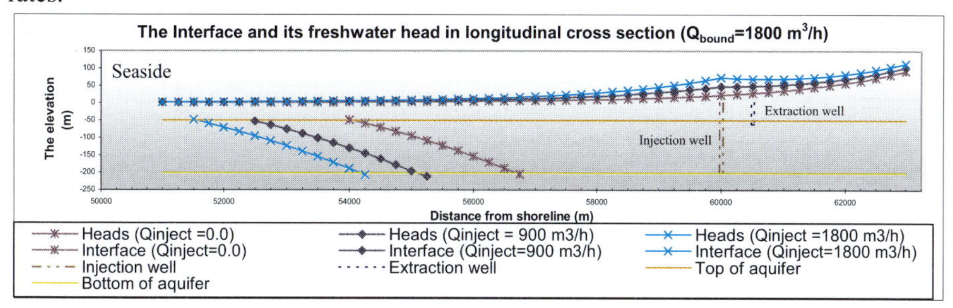

Figure 3.6: The responses of the freshwater head and interface to the injection rates.

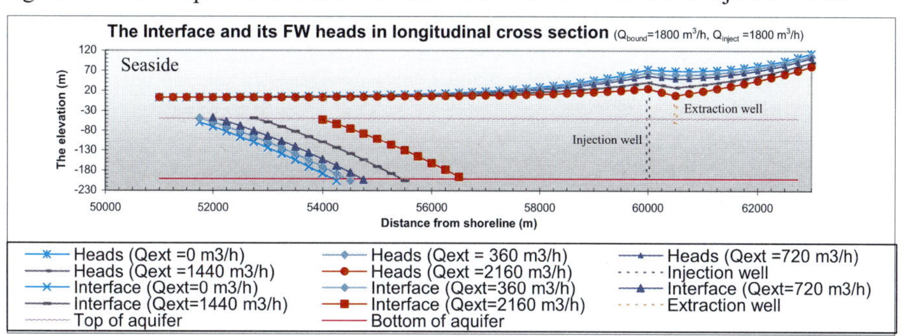

Figure 3.9: The responses of freshwater head and interface to the extraction rates.

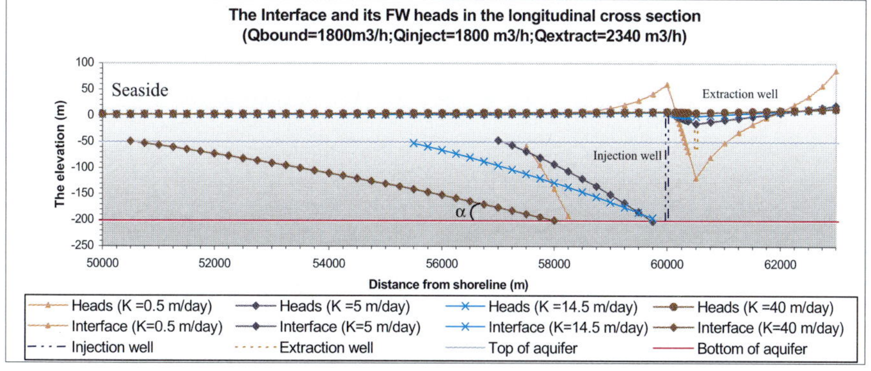

Figure 3.12: The responses of freshwater head and interface to the changes of the hydraulic conductivity.

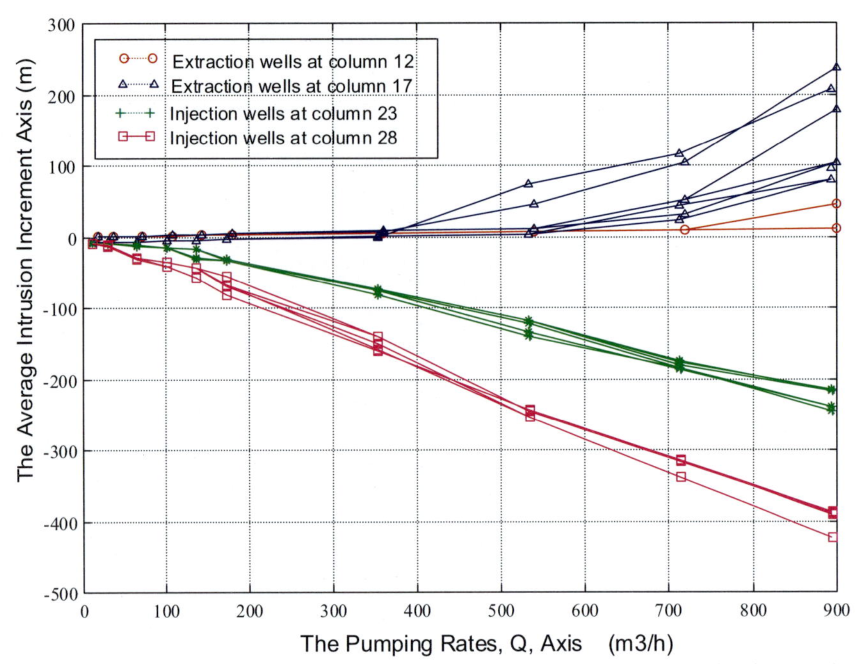

Figure 3.15: The average intrusion increments and pumping rates, Q, in the quasi-three-dimensional model. (The wells at columns 12, 17, 23, 28 are far from the shoreline with the distances of 59500 m, 58250 m, 56750 m and 55500 m respectively.)

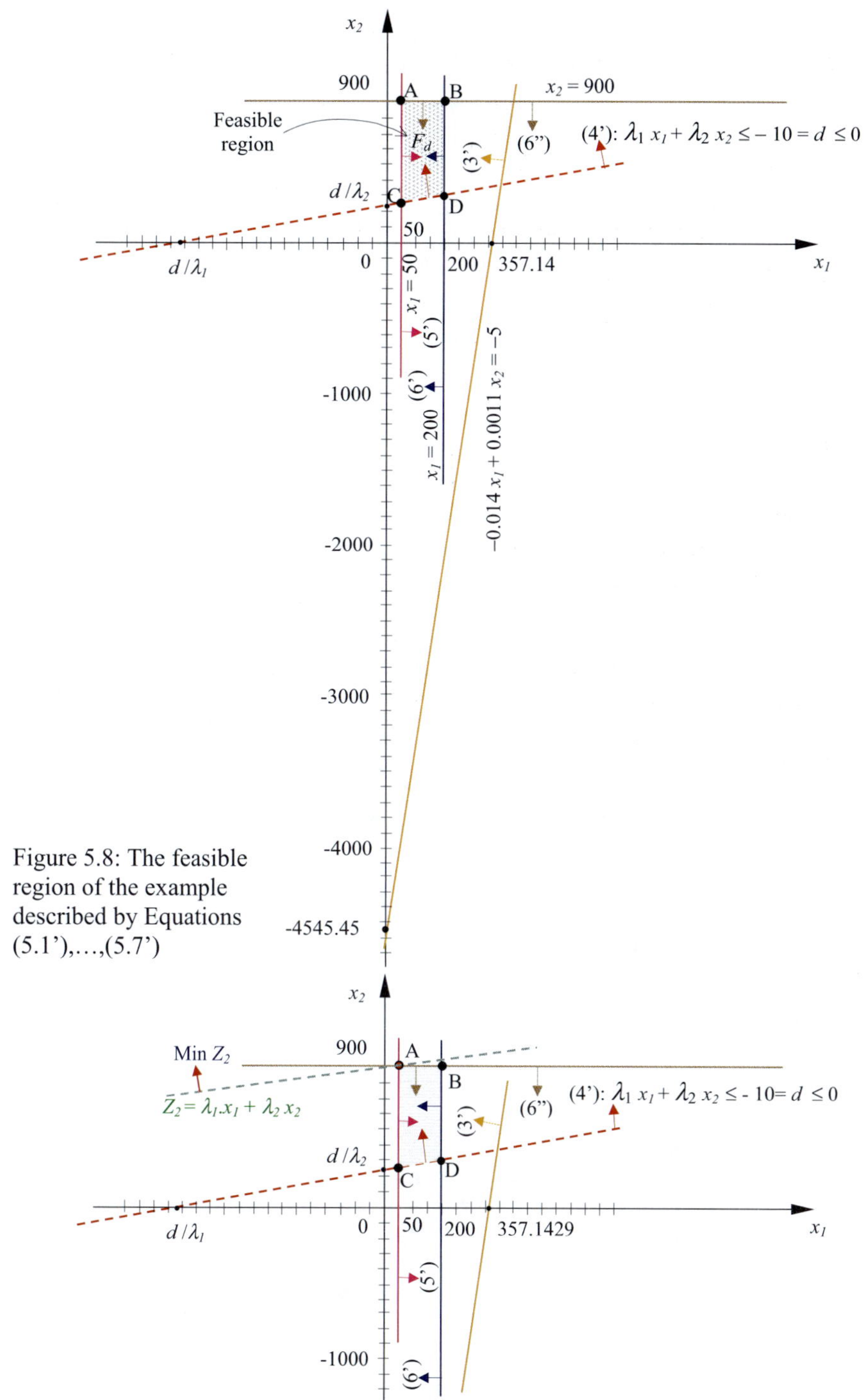

Figure 5.8: The feasible region of the example described by Equations (5.1’),…,(5.7’)

Figure 5.9: Graphical representation of the second objective function and its optimal solution at point A.

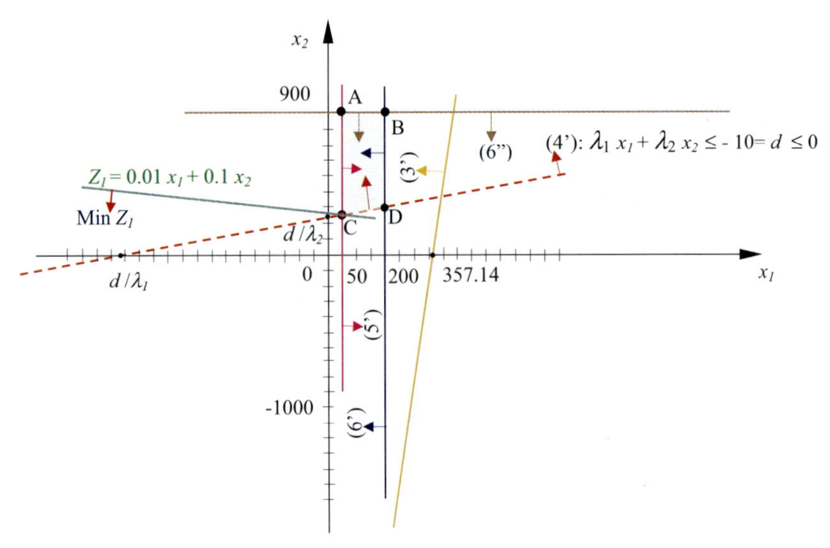

Figure 5.10: Graphical representation of the first objective function and its optimal solution at point C.

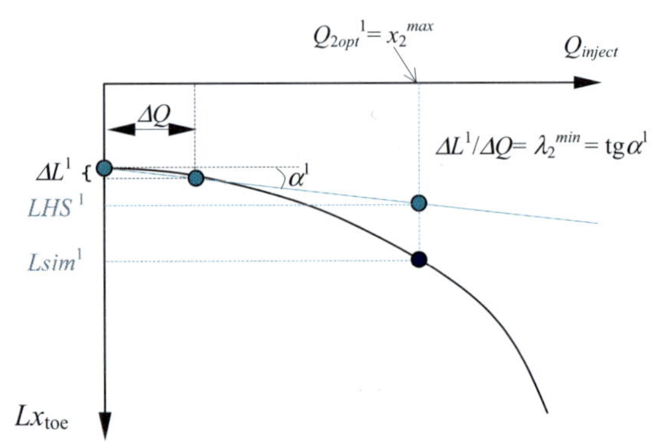

Figure 5.12: Schematic diagram of the iterative solution for finding the optimal solution.

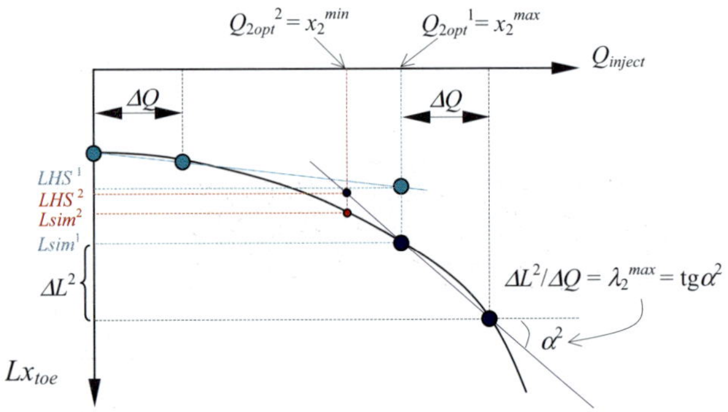

Figure 5.13: Schematic diagram of the iterative solution for finding the optimal solution continued

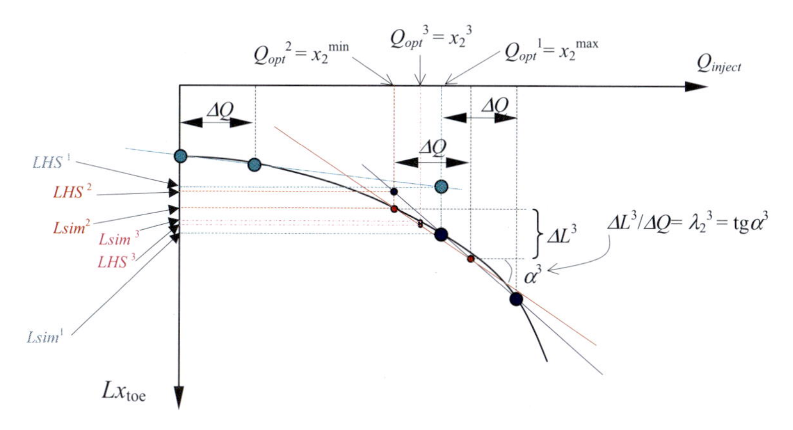

Figure 5.14: Schematic diagram of the iterative solution for finding the optimal solution continued.

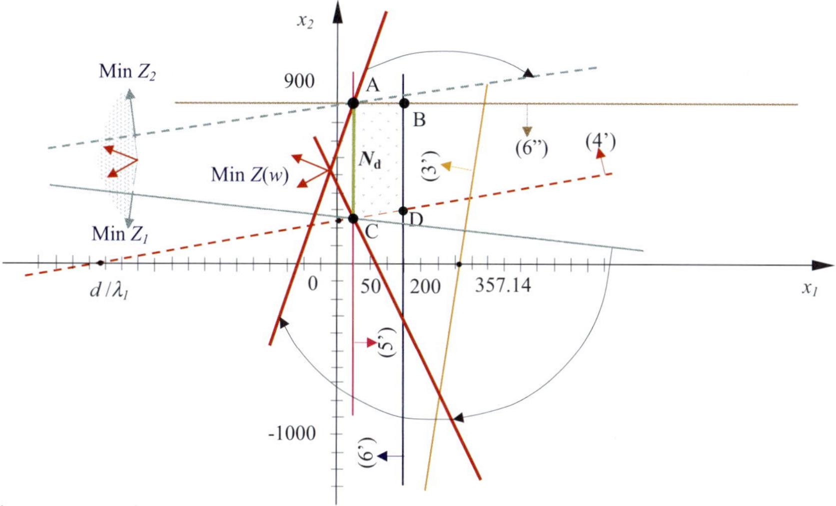

Figure 5.15: The feasible region and the multi-objective functions by the weighting method.

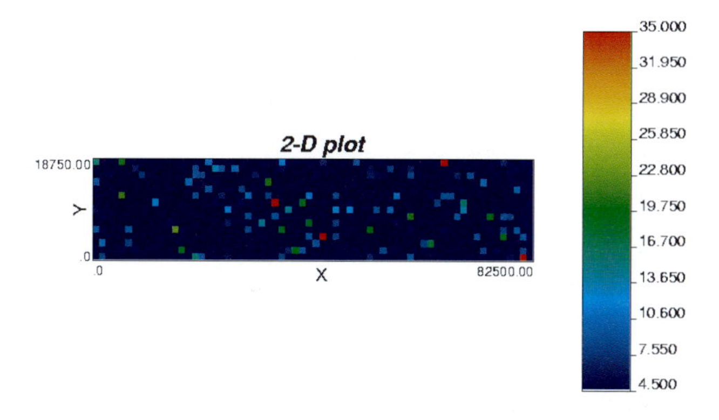

Figure 5.18. The 2-D plan view for the first realization of hydraulic conductivity.

```
MAP OF EXTENT OF INTRUSION (F-FRESHWATER, M-FRESH AND SALTWATER, S-SALTWATER)
                        10        20        30        40        50        60       146
              1234567890123456789012345678901234567890134567890123456789012345  12456
      15      ................................................................  .....
      14      .FRRRRRRRRREFFFFEFFFFFIFFFFIMMMMMMMMMSSSSSSSSSSSSSSSSSSSSSSSSSSS  SSSSS
      13      .FRRRRRRRRRFFFFFFFFFFFFFFFFFMMMMMMMMMSSSSSSSSSSSSSSSSSSSSSSSSSSS  SSSSS
      12      .FRRRRRRRRREFFFFEFFFFFIFFFFIMMMMMMMMMSSSSSSSSSSSSSSSSSSSSSSSSSSS  SSSSS
      11      .FRRRRRRRRRFFFFFFFFFFFFFFFFFMMMMMMMMMSSSSSSSSSSSSSSSSSSSSSSSSSSS  SSSSS
      10      .FRRRRRRRRREFFFFEFFFFFIFFFFIMMMMMMMMMSSSSSSSSSSSSSSSSSSSSSSSSSSS  SSSSS
       9      .FRRRRRRRRRFFFFFFFFFFFFFFFFFMMMMMMMMMSSSSSSSSSSSSSSSSSSSSSSSSSSS  SSSSS
       8      .FRRRRRRRRREFFFFEFFFFFIFFFFIMMMMMMMMMSSSSSSSSSSSSSSSSSSSSSSSSSS   SSSSS
       7      .FRRRRRRRRRFFFFFFFFFFFFFFFFFMMMMMMMMMSSSSSSSSSSSSSSSSSSSSSSSSSS  SSSSSS
       6      .FRRRRRRRRREFFFFEFFFFFIFFFFIMMMMMMMMMSSSSSSSSSSSSSSSSSSSSSSSSSS  SSSSSS
       5      .FRRRRRRRRRFFFFFFFFFFFFFFFFFMMMMMMMMMSSSSSSSSSSSSSSSSSSSSSSSSSS  SSSSSS
       4      .FRRRRRRRRREFFFFEFFFFFIFFFFIMMMMMMMMMSSSSSSSSSSSSSSSSSSSSSSSSSS  SSSSSS
       3      .FRRRRRRRRRFFFFFFFFFFFFFFFFFMMMMMMMMMSSSSSSSSSSSSSSSSSSSSSSSSSS  SSSSSS
       2      .FRRRRRRRRREFFFFEFFFFFIFFFFIMMMMMMMMMSSSSSSSSSSSSSSSSSSSSSSSSSS  SSSSSS
       1      ................................................................  .....
              1234567890123456789012345678901234567890134567890123456789012345  23456
                        10        20        30        40        50        60       146
(NOTE: R-RESIDENTIAL AREA. E- CANDIDATE EXTRACTION. I- CANDIDATE INJECTION)
```

Figure 6.5: The SW toe and FW tip at the non-pumping condition.

```
MAP OF EXTENT OF INTRUSION (F-FRESHWATER, M-FRESH AND SALTWATER, S-SALTWATER)
                        10        20        30        40        50        60       146
              1234567890123456789012345678901234567890134567890123456789012345  12456
      15      ................................................................  .....
      14      .FRRRRRRRRREFFFFEFFFFFIFFFFIFFFFFFFFFFFMMMMMMMMMMMSSSSSSSSSSSSSSS  SSSSS
      13      .FRRRRRRRRRFFFFFFFFFFFFFFFFFFFFFFFFFFFFMMMMMMMMMMMSSSSSSSSSSSSSSS  SSSSS
      12      .FRRRRRRRRREFFFFEFFFFFIFFFFIFFFFFFFFFFFMMMMMMMMMMSSSSSSSSSSSSSSSS   SSSS
      11      .FRRRRRRRRRFFFFFFFFFFFFFFFFFFFFFFFFFFFMMMMMMMMMMMSSSSSSSSSSSSSSSS  SSSSS
      10      .FRRRRRRRRREFFFFEFFFFFIFFFFIFFFFFFFFFFMMMMMMMMMMSSSSSSSSSSSSSSSSS  SSSSS
       9      .FRRRRRRRRRFFFFFFFFFFFFFFFFFFFFFFFFFFFMMMMMMMMMMSSSSSSSSSSSSSSSS   SSSSS
       8      .FRRRRRRRRREFFFFEFFFFFIFFFFIFFFFFFFFFFMMMMMMMMMMSSSSSSSSSSSSSSSS   SSSSS
       7      .FRRRRRRRRRFFFFFFFFFFFFFFFFFFFFFFFFFFFMMMMMMMMMMSSSSSSSSSSSSSSSS  SSSSSS
       6      .FRRRRRRRRREFFFFEFFFFFIFFFFIFFFFFFFFFFMMMMMMMMMMSSSSSSSSSSSSSSSS  SSSSSS
       5      .FRRRRRRRRRFFFFFFFFFFFFFFFFFFFFFFFFFFFFMMMMMMMMMMMSSSSSSSSSSSSSSS  SSSSSS
       4      .FRRRRRRRRREFFFFEFFFFFIFFFFIFFFFFFFFFFFMMMMMMMMMMMSSSSSSSSSSSSSSS  SSSSSS
       3      .FRRRRRRRRRFFFFFFFFFFFFFFFFFFFFFFFFFFFFFMMMMMMMMMMMSSSSSSSSSSSSSS  SSSSSS
       2      .FRRRRRRRRREFFFFEFFFFFIFFFFIFFFFFFFFFFFMMMMMMMMMMMMSSSSSSSSSSSSSS  SSSSSS
       1      ......................................................................
              1234567890123456789012345678901234567890134567890123456789012345  23456
                        10        20        30        40        50        60       146
NOTE: R-RESIDENTIAL AREA, E-CANDIDATE EXTRACTION, I- CANDIDATE INJECTION, E-OPTIMAL EXTRACTION, I-OPTIMAL INJECTION
```

Figure 6.6: The SW toe and FW tip after the first iteration.

```
MAP OF EXTENT OF INTRUSION (F-FRESHWATER, M-FRESH AND SALTWATER, S-SALTWATER)
                        10        20        30        40        50        60       146
              1234567890123456789012345678901234567890134567890123456789012345  12456
      15      ................................................................  .....
      14      .FRRRRRRRRREFFFFEFFFFFIFFFFIFFFFFFFFFFFMMMMMMMMMMMSSSSSSSSSSSSSSS  SSSSS
      13      .FRRRRRRRRRFFFFFFFFFFFFFFFFFFFFFFFFFFFFMMMMMMMMMMMSSSSSSSSSSSSSSS  SSSSS
      12      .FRRRRRRRRREFFFFEFFFFFIFFFFIFFFFFFFFFFFMMMMMMMMMMSSSSSSSSSSSSSSSS   SSSS
      11      .FRRRRRRRRRFFFFFFFFFFFFFFFFFFFFFFFFFFFMMMMMMMMMMMSSSSSSSSSSSSSSSS  SSSSS
      10      .FRRRRRRRRREFFFFEFFFFFIFFFFIFFFFFFFFFFMMMMMMMMMMSSSSSSSSSSSSSSSSS  SSSSS
       9      .FRRRRRRRRRFFFFFFFFFFFFFFFFFFFFFFFFFFFMMMMMMMMMMSSSSSSSSSSSSSSSS   SSSSS
       8      .FRRRRRRRRREFFFFEFFFFFIFFFFIFFFFFFFFFFMMMMMMMMMMSSSSSSSSSSSSSSSS   SSSSS
       7      .FRRRRRRRRRFFFFFFFFFFFFFFFFFFFFFFFFFFFMMMMMMMMMMSSSSSSSSSSSSSSSS  SSSSSS
       6      .FRRRRRRRRREFFFFEFFFFFIFFFFIFFFFFFFFFFMMMMMMMMMMSSSSSSSSSSSSSSSS  SSSSSS
       5      .FRRRRRRRRRFFFFFFFFFFFFFFFFFFFFFFFFFFFFMMMMMMMMMMMSSSSSSSSSSSSSSS  SSSSSS
       4      .FRRRRRRRRREFFFFEFFFFFIFFFFIFFFFFFFFFFFMMMMMMMMMMMSSSSSSSSSSSSSSS  SSSSSS
       3      .FRRRRRRRRRFFFFFFFFFFFFFFFFFFFFFFFFFFFFFMMMMMMMMMMMSSSSSSSSSSSSSS  SSSSSS
       2      .FRRRRRRRRREFFFFEFFFFFIFFFFIFFFFFFFFFFFMMMMMMMMMMMMSSSSSSSSSSSSSS  SSSSSS
       1      ......................................................................
              1234567890123456789012345678901234567890134567890123456789012345  23456
                        10        20        30        40        50        60       146
NOTE: R-RESIDENTIAL AREA, E-CANDIDATE EXTRACTION, I- CANDIDATE INJECTION, E-OPTIMAL EXTRACTION, I-OPTIMAL INJECTION
```

Figure 6.7: The SW toe and FW tip after the third iteration.

```
MAP OF EXTENT OF INTRUSION (F-FRESHWATER, M-FRESH AND SALTWATER, S-SALTWATER)
                     10        20        30        40        50        60        146
            1234567890123456789012345678901234567890134567890123456789012345  12456
   15       ................................................................  .....
   14       .FRRRRRRRRREFFFFEFFFFFIFFFFIMMMMMMMMMMSSSSSSSSSSSSSSSSSSSSSSSSSSS  SSSSS
   13       .FRRRRRRRRRFFFFFFFFFFFFFFFFFMMMMMMMMMMSSSSSSSSSSSSSSSSSSSSSSSSSSS  SSSSS
   12       .FRRRRRRRRREFFFFEFFFFFIFFFFIMMMMMMMMMMSSSSSSSSSSSSSSSSSSSSSSSSSSS  SSSSS
   11       .FRRRRRRRRRFFFFFFFFFFFFFFFFFMMMMMMMMMMSSSSSSSSSSSSSSSSSSSSSSSSSSS  SSSSS
   10       .FRRRRRRRRREFFFFEFFFFFIFFFFIMMMMMMMMMMSSSSSSSSSSSSSSSSSSSSSSSSSSS  SSSSS
    9       .FRRRRRRRRRFFFFFFFFFFFFFFFFFMMMMMMMMMMSSSSSSSSSSSSSSSSSSSSSSSSSSS  SSSSS
    8       .FRRRRRRRRREFFFFEFFFFFIFFFFIMMMMMMMMMMSSSSSSSSSSSSSSSSSSSSSSSSSSS  SSSSS
    7       .FRRRRRRRRRFFFFFFFFFFFFFFFFFMMMMMMMMMMSSSSSSSSSSSSSSSSSSSSSSSSSSS  SSSSS
    6       .FRRRRRRRRREFFFFEFFFFFIFFFFIMMMMMMMMMMSSSSSSSSSSSSSSSSSSSSSSSSSSS  SSSSS
    5       .FRRRRRRRRRFFFFFFFFFFFFFFFFFMMMMMMMMMMSSSSSSSSSSSSSSSSSSSSSSSSSSS  SSSSS
    4       .FRRRRRRRRREFFFFEFFFFFIFFFFIMMMMMMMMMMSSSSSSSSSSSSSSSSSSSSSSSSSSS  SSSSS
    3       .FRRRRRRRRRFFFFFFFFFFFFFFFFFMMMMMMMMMMSSSSSSSSSSSSSSSSSSSSSSSSSSS  SSSSS
    2       .FRRRRRRRRREFFFFEFFFFFIFFFFIMMMMMMMMMMSSSSSSSSSSSSSSSSSSSSSSSSSSS  SSSSS
    1       ................................................................  .....
            1234567890123456789012345678901234567890134567890123456789012345  23456
                     10        20        30        40        50        60        146
(NOTE: R-RESIDENTIAL AREA. E- CANDIDATE EXTRACTION. I- CANDIDATE INJECTION)
```

Figure 6.10: The SW toe and FW tip at the non-pumping condition.

```
MAP OF EXTENT OF INTRUSION (F-FRESHWATER, M-FRESH AND SALTWATER, S-SALTWATER)
                     10        20        30        40        50        60        146
            1234567890123456789012345678901234567890134567890123456789012345  12456
   15       ................................................................  .....
   14       .FRRRRRRRRREFFFFEFFFFFIFFFFIFFFFFFFFFFMMMMMMMMMMSSSSSSSSSSSSSSSSS  SSSSS
   13       .FRRRRRRRRRFFFFFFFFFFFFFFFFFFFFFFFFFMMMMMMMMMMMSSSSSSSSSSSSSSSSSS  SSSSS
   12       .FRRRRRRRRREFFFFEFFFFFIFFFFIFFFFFFFMMMMMMMMMMSSSSSSSSSSSSSSSSSSSS  SSSS
   11       .FRRRRRRRRRFFFFFFFFFFFFFFFFFFFFFMMMMMMMMMMMMSSSSSSSSSSSSSSSSSSSSS  SSSSS
   10       .FRRRRRRRRREFFFFEFFFFFIFFFFIFMMMMMMMMMMMMMSSSSSSSSSSSSSSSSSSSSSSS  SSSSS
    9       .FRRRRRRRRRFFFFFFFFFFFFFFFFFMMMMMMMMMMMMSSSSSSSSSSSSSSSSSSSSSSSSS  SSSSS
    8       .FRRRRRRRRREFFFFEFFFFFIFFFFIMMMMMMMMMMMMSSSSSSSSSSSSSSSSSSSSSSSSS  SSSSSS
    7       .FRRRRRRRRRFFFFFFFFFFFFFFFFFMMMMMMMMMMMMMMSSSSSSSSSSSSSSSSSSSSSSS  SSSSSS
    6       .FRRRRRRRRREFFFFEFFFFFIFFFFIFFFMMMMMMMMMMMMMSSSSSSSSSSSSSSSSSSSSS  SSSSSS
    5       .FRRRRRRRRRFFFFFFFFFFFFFFFFFFFFFFFFMMMMMMMMMMMMSSSSSSSSSSSSSSSSSS  SSSSSS
    4       .FRRRRRRRRREFFFFEFFFFFIFFFFIFFFFFFFFFMMMMMMMMMMMSSSSSSSSSSSSSSSSS  SSSSSS
    3       .FRRRRRRRRRFFFFFFFFFFFFFFFFFFFFFFFFFFFMMMMMMMMMMMSSSSSSSSSSSSSSSS  SSSSSS
    2       .FRRRRRRRRREFFFFEFFFFFIFFFFIFFFFFFFFFFFMMMMMMMMMMMSSSSSSSSSSSSSSS  SSSSS
    1       ......................................................................
            1234567890123456789012345678901234567890134567890123456789012345  23456
                     10        20        30        40        50        60        146
NOTE: R-RESIDENTIAL AREA, E-CANDIDATE EXTRACTION, I- CANDIDATE INJECTION, E-OPTIMAL EXTRACTION, I-OPTIMAL INJECTION
```

Figure 6.11: The SW toe and FW tip after the first iteration.

```
MAP OF EXTENT OF INTRUSION (F-FRESHWATER, M-FRESH AND SALTWATER, S-SALTWATER)
                     10        20        30        40        50        60        146
            1234567890123456789012345678901234567890134567890123456789012345  12456
   15       ................................................................  .....
   14       .FRRRRRRRRREFFFFEFFFFFIFFFFIFFFFFFFFFFFMMMMMMMMMMSSSSSSSSSSSSSSSS  SSSSS
   13       .FRRRRRRRRRFFFFFFFFFFFFFFFFFFFFFFFFFFFMMMMMMMMMMSSSSSSSSSSSSSSSSS  SSSSS
   12       .FRRRRRRRRREFFFFEFFFFFIFFFFIFFFFFFFFFMMMMMMMMMMSSSSSSSSSSSSSSSSSS  SSSS
   11       .FRRRRRRRRRFFFFFFFFFFFFFFFFFFFFFFFFMMMMMMMMMMMSSSSSSSSSSSSSSSSSSS  SSSSS
   10       .FRRRRRRRRREFFFFEFFFFFIFFFFIFFFMMMMMMMMMMMMMSSSSSSSSSSSSSSSSSSSSS  SSSSS
    9       .FRRRRRRRRRFFFFFFFFFFFFFFFFFFMMMMMMMMMMMMMSSSSSSSSSSSSSSSSSSSSSSS  SSSSS
    8       .FRRRRRRRRREFFFFEFFFFFIFFFFIMMMMMMMMMMMMSSSSSSSSSSSSSSSSSSSSSSSSS  SSSSSS
    7       .FRRRRRRRRRFFFFFFFFFFFFFFFFFFMMMMMMMMMMMMMSSSSSSSSSSSSSSSSSSSSSSS  SSSSSS
    6       .FRRRRRRRRREFFFFEFFFFFIFFFFIFFFMMMMMMMMMMMMMSSSSSSSSSSSSSSSSSSSSSS  SSSSSS
    5       .FRRRRRRRRRFFFFFFFFFFFFFFFFFFFFFFFFMMMMMMMMMMMMSSSSSSSSSSSSSSSSSS  SSSSSS
    4       .FRRRRRRRRREFFFFEFFFFFIFFFFIFFFFFFFFFMMMMMMMMMMMSSSSSSSSSSSSSSSSS  SSSSSS
    3       .FRRRRRRRRRFFFFFFFFFFFFFFFFFFFFFFFFFFFMMMMMMMMMMMSSSSSSSSSSSSSSSS  SSSSSS
    2       .FRRRRRRRRREFFFFEFFFFFIFFFFIFFFFFFFFFFFMMMMMMMMMMMSSSSSSSSSSSSSSS  SSSSS
    1       ......................................................................
            1234567890123456789012345678901234567890134567890123456789012345  23456
                     10        20        30        40        50        60        146
NOTE: R-RESIDENTIAL AREA, E-CANDIDATE EXTRACTION, I- CANDIDATE INJECTION, E-OPTIMAL EXTRACTION, I-OPTIMAL INJECTION
```

Figure 6.12: The SW toe and FW tip after the third iteration.

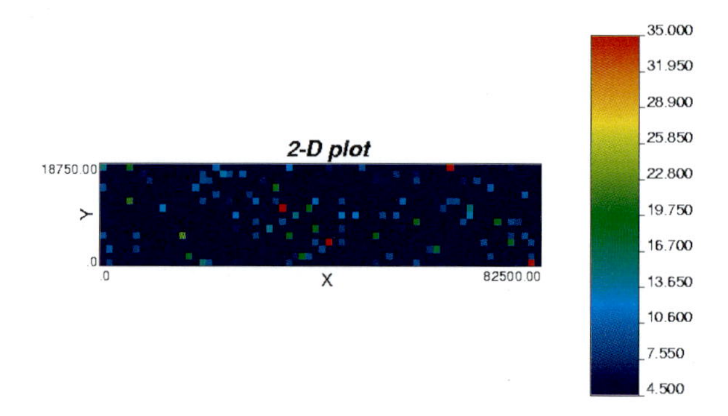

Figure 6.13: The 2-D distribution of the first realization of hydraulic conductivity.

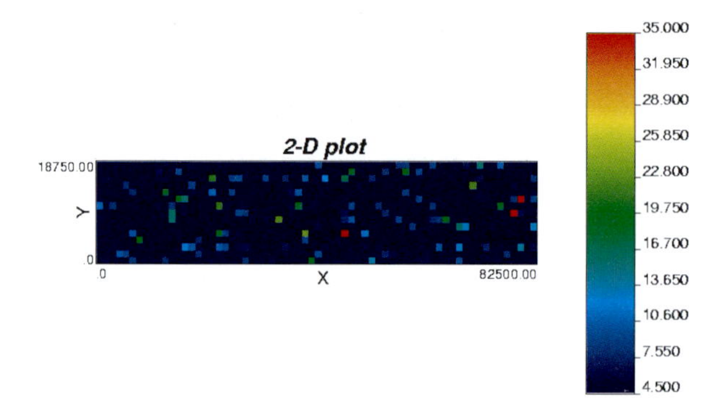

Figure 6.14: The 2-D distribution of the twentieth realization of hydraulic conductivity.

```
MAP OF EXTENT OF INTRUSION (F-FRESHWATER. M-FRESH AND SALTWATER. S-SALTWATER)
                    10        20        30        40        50        60       146
          1234567890123456789012345678901234567890134567890123456789012345  12456
    15    ................................................................  .....
    14    .FRRRRRRRRRREFFFFEFFFFFIFFFFIMMMMMMMMSSSSSSSSSSSSSSSSSSSSSSSSSSSS  SSSSS
    13    .FRRRRRRRRRFFFFFFFFFFFFFFFFFMMMMMMMMMSSSSSSSSSSSSSSSSSSSSSSSSSSSS  SSSSS
    12    .FRRRRRRRRRREFFFFEFFFFFIFFFFIMMMMMMMMMSSSSSSSSSSSSSSSSSSSSSSSSSSS  SSSSS
    11    .FRRRRRRRRRFFFFFFFFFFFFFFFFFMMMMMMMMMSSSSSSSSSSSSSSSSSSSSSSSSSSSS  SSSSS
    10    .FRRRRRRRRRREFFFFEFFFFFIFFFFIMMMMMMMMMSSSSSSSSSSSSSSSSSSSSSSSSSSS  SSSSS
     9    .FRRRRRRRRRFFFFFFFFFFFFFFFFFMMMMMMMMMSSSSSSSSSSSSSSSSSSSSSSSSSSS   SSSSS
     8    .FRRRRRRRRRREFFFFEFFFFFIFFFFIMMMMMMMMMSSSSSSSSSSSSSSSSSSSSSSSSSS   SSSSS
     7    .FRRRRRRRRRFFFFFFFFFFFFFFFFFMMMMMMMMMSSSSSSSSSSSSSSSSSSSSSSSSSSS  SSSSSS
     6    .FRRRRRRRRRREFFFFEFFFFFIFFFFIMMMMMMMMMSSSSSSSSSSSSSSSSSSSSSSSSSS  SSSSSS
     5    .FRRRRRRRRRFFFFFFFFFFFFFFFFFMMMMMMMMMSSSSSSSSSSSSSSSSSSSSSSSSSS   SSSSSS
     4    .FRRRRRRRRRREFFFFEFFFFFIFFFFIMMMMMMMMSSSSSSSSSSSSSSSSSSSSSSSSSS   SSSSSS
     3    .FRRRRRRRRRFFFFFFFFFFFFFFFFFMMMMMMMMSSSSSSSSSSSSSSSSSSSSSSSSSSS   SSSSSS
     2    .FRRRRRRRRRREFFFFEFFFFFIFFFFIMMMMMMMMMSSSSSSSSSSSSSSSSSSSSSSSSSS  SSSSSS
     1    ................................................................  .....
          1234567890123456789012345678901234567890134567890123456789012345  23456
                    10        20        30        40        50        60       146
(NOTE: R-RESIDENTIAL AREA. E- CANDIDATE EXTRACTION. I- CANDIDATE INJECTION)
```

Figure 6.18: The SW toe and FW tip at the non-pumping condition.

```
MAP OF EXTENT OF INTRUSION (F-FRESHWATER. M-FRESH AND SALTWATER. S-SALTWATER)
                  10        20        30        40       50        60         146
         1234567890123456789012345678901234567890134567890123456789012345  12456
   15    .................................................................  .....
   14    .FRRRRRRRRREFFFFEFFFFFIFFFFIFFFFFFFFMMMMMMMMMSSSSSSSSSSSSSSSSSSSS  SSSSS
   13    .FRRRRRRRRRFFFFFFFFFFFFFFFFFFFFFFFFMMMMMMMMMSSSSSSSSSSSSSSSSSSSSS  SSSSS
   12    .FRRRRRRRRREFFFFEFFFFFIFFFFIFFFFFFFMMMMMMMMMSSSSSSSSSSSSSSSSSSSSS  SSSSS
   11    .FRRRRRRRRRFFFFFFFFFFFFFFFFFFFFFFMMMMMMMMMMSSSSSSSSSSSSSSSSSSSSSS  SSSSS
   10    .FRRRRRRRRREFFFFEFFFFFIFFFFIFFFFFMMMMMMMMMSSSSSSSSSSSSSSSSSSSSSSS  SSSSS
    9    .FRRRRRRRRRFFFFFFFFFFFFFFFFFFFFFFFMMMMMMMMMSSSSSSSSSSSSSSSSSSSSSS  SSSSS
    8    .FRRRRRRRRREFFFFEFFFFFIFFFFIFFFFFFFMMMMMMMMMSSSSSSSSSSSSSSSSSSSSS  SSSSS
    7    .FRRRRRRRRRFFFFFFFFFFFFFFFFFFFFFFFMMMMMMMMMMSSSSSSSSSSSSSSSSSSSSS  SSSSS
    6    .FRRRRRRRRREFFFFEFFFFFIFFFFIFFFFFFFMMMMMMMMMSSSSSSSSSSSSSSSSSSSSS  SSSSS
    5    .FRRRRRRRRRFFFFFFFFFFFFFFFFFFFFFFFFMMMMMMMMMSSSSSSSSSSSSSSSSSSSSS  SSSSS
    4    .FRRRRRRRRREFFFFEFFFFFIFFFFIFFFFFFFMMMMMMMMMMSSSSSSSSSSSSSSSSSSSS  SSSSS
    3    .FRRRRRRRRRFFFFFFFFFFFFFFFFFFFFFFFFMMMMMMMMMMSSSSSSSSSSSSSSSSSSSS  SSSSS
    2    .FRRRRRRRRREFFFFEFFFFFIFFFFIFFFFFFFFFMMMMMMMMSSSSSSSSSSSSSSSSSSSS  SSSSS
    1    .................................................................  .....
         1234567890123456789012345678901234567890134567890123456789012345  23456
                  10        20        30        40       50        60         146
NOTE: R-RESIDENTIAL AREA, E-CANDIDATE EXTRACTION, I- CANDIDATE INJECTION, E-OPTIMAL EXTRACTION, I-OPTIMAL INJECTION
```

Figure 6.19: The SW toe and FW tip after the first iteration.

```
MAP OF EXTENT OF INTRUSION (F-FRESHWATER. M-FRESH AND SALTWATER. S-SALTWATER)
                  10        20        30        40       50        60         146
         1234567890123456789012345678901234567890134567890123456789012345  12456
   15    .................................................................  .....
   14    .FRRRRRRRRREFFFFEFFFFFIFFFFIFFFFFFFFMMMMMMMMMSSSSSSSSSSSSSSSSSSSS  SSSSS
   13    .FRRRRRRRRRFFFFFFFFFFFFFFFFFFFFFFFFMMMMMMMMMMSSSSSSSSSSSSSSSSSSSS  SSSSS
   12    .FRRRRRRRRREFFFFEFFFFFIFFFFIFFFFFFFMMMMMMMMMMSSSSSSSSSSSSSSSSSSSS  SSSSS
   11    .FRRRRRRRRRFFFFFFFFFFFFFFFFFFFFFFFFMMMMMMMMMSSSSSSSSSSSSSSSSSSSSS  SSSSS
   10    .FRRRRRRRRREFFFFEFFFFFIFFFFIFFFFFFFMMMMMMMMMSSSSSSSSSSSSSSSSSSSSS  SSSSS
    9    .FRRRRRRRRRFFFFFFFFFFFFFFFFFFFFFFFFMMMMMMMMMSSSSSSSSSSSSSSSSSSSSS  SSSSS
    8    .FRRRRRRRRREFFFFEFFFFFIFFFFIFFFFFFFFMMMMMMMMSSSSSSSSSSSSSSSSSSSSS  SSSSS
    7    .FRRRRRRRRRFFFFFFFFFFFFFFFFFFFFFFFFMMMMMMMMMMSSSSSSSSSSSSSSSSSSSS  SSSSS
    6    .FRRRRRRRRREFFFFEFFFFFIFFFFIFFFFFFFMMMMMMMMMSSSSSSSSSSSSSSSSSSSSS  SSSSS
    5    .FRRRRRRRRRFFFFFFFFFFFFFFFFFFFFFFFFMMMMMMMMMSSSSSSSSSSSSSSSSSSSSS  SSSSS
    4    .FRRRRRRRRREFFFFEFFFFFIFFFFIFFFFFFFMMMMMMMMMMSSSSSSSSSSSSSSSSSSSS  SSSSS
    3    .FRRRRRRRRRFFFFFFFFFFFFFFFFFFFFFFFFMMMMMMMMMMSSSSSSSSSSSSSSSSSSSS  SSSSS
    2    .FRRRRRRRRREFFFFEFFFFFIFFFFIFFFFFFFFFMMMMMMMMSSSSSSSSSSSSSSSSSSSS  SSSSS
    1    .................................................................  .....
         1234567890123456789012345678901234567890134567890123456789012345  23456
                  10        20        30        40       50        60         146
NOTE: R-RESIDENTIAL AREA, E-CANDIDATE EXTRACTION, I- CANDIDATE INJECTION, E-OPTIMAL EXTRACTION, I-OPTIMAL INJECTION
```

Figure 6.20: The SW toe and FW tip after the ninth iteration.

```
MAP OF EXTENT OF INTRUSION (F-FRESHWATER. M-FRESH AND SALTWATER. S-SALTWATER)
                  10        20        30        40       50        60         146
         1234567890123456789012345678901234567890134567890123456789012345  12456
   15    .................................................................  .....
   14    .FRRRRRRRRREFFFFEFFFFFIFFFFIMMMMMMMMSSSSSSSSSSSSSSSSSSSSSSSSSSSSS  SSSSS
   13    .FRRRRRRRRRFFFFFFFFFFFFFFFFFMMMMMMMMMSSSSSSSSSSSSSSSSSSSSSSSSSSSS  SSSSS
   12    .FRRRRRRRRREFFFFEFFFFFIFFFFIMMMMMMMMMSSSSSSSSSSSSSSSSSSSSSSSSSSSS  SSSSS
   11    .FRRRRRRRRRFFFFFFFFFFFFFFFFFMMMMMMMMMSSSSSSSSSSSSSSSSSSSSSSSSSSSS  SSSSS
   10    .FRRRRRRRRREFFFFEFFFFFIFFFFIMMMMMMMMMSSSSSSSSSSSSSSSSSSSSSSSSSSSS  SSSSS
    9    .FRRRRRRRRRFFFFFFFFFFFFFFFFFMMMMMMMMMSSSSSSSSSSSSSSSSSSSSSSSSSSSS  SSSSS
    8    .FRRRRRRRRREFFFFEFFFFFIFFFFIMMMMMMMMMSSSSSSSSSSSSSSSSSSSSSSSSSSSS  SSSSS
    7    .FRRRRRRRRRFFFFFFFFFFFFFFFFFMMMMMMMMMSSSSSSSSSSSSSSSSSSSSSSSSSSSS  SSSSS
    6    .FRRRRRRRRREFFFFEFFFFFIFFFFIMMMMMMMMMSSSSSSSSSSSSSSSSSSSSSSSSSSSS  SSSSS
    5    .FRRRRRRRRRFFFFFFFFFFFFFFFFFMMMMMMMMMSSSSSSSSSSSSSSSSSSSSSSSSSSSS  SSSSS
    4    .FRRRRRRRRREFFFFEFFFFFIFFFFIMMMMMMMMSSSSSSSSSSSSSSSSSSSSSSSSSSSSS  SSSSS
    3    .FRRRRRRRRRFFFFFFFFFFFFFFFFFMMMMMMMMSSSSSSSSSSSSSSSSSSSSSSSSSSSSS  SSSSS
    2    .FRRRRRRRRREFFFFEFFFFFIFFFFIMMMMMMMMMSSSSSSSSSSSSSSSSSSSSSSSSSSSS  SSSSS
    1    .................................................................  .....
         1234567890123456789012345678901234567890134567890123456789012345  23456
                  10        20        30        40       50        60         146
NOTE: R-RESIDENTIAL AREA. E- CANDIDATE EXTRACTION. I- CANDIDATE INJECTION
```

Figure 6.23: The SW toe and FW tip at the non-pumping condition simulated with the twentieth realization of the hydraulic.

```
   MAP OF EXTENT OF INTRUSION (F-FRESHWATER. M-FRESH AND SALTWATER. S-SALTWATER)
                 10        20        30        40        50        60         146
        12345678901234567890123456789012345678901345678901234567890123456789012345  12456
   15   ..................................................................  .....
   14   .FRRRRRRRRRE FFFFEFFFFFIFFFFIFFFFFFFFFMMMMMMMMMSSSSSSSSSSSSSSSSSSSS  SSSSS
   13   .FRRRRRRRRRFFFFFFFFFFFFFFFFFFFFFFFFMMMMMMMMMSSSSSSSSSSSSSSSSSSSSS  SSSSS
   12   .FRRRRRRRRREFFFFEFFFFFIFFFFIFFFFFFFMMMMMMMMMSSSSSSSSSSSSSSSSSSSSS  SSSSS
   11   .FRRRRRRRRRFFFFFFFFFFFFFFFFFFFFFMMMMMMMMMMSSSSSSSSSSSSSSSSSSSSSSS  SSSSS
   10   .FRRRRRRRRREFFFFEFFFFFIFFFFIFMMMMMMMMMMMSSSSSSSSSSSSSSSSSSSSSSSSS  SSSSS
    9   .FRRRRRRRRRFFFFFFFFFFFFFFFFFMMMMMMMMMMSSSSSSSSSSSSSSSSSSSSSSSSSSS  SSSSS
    8   .FRRRRRRRRREFFFFEFFFFFIFFFFIFMMMMMMMMMMMSSSSSSSSSSSSSSSSSSSSSSSSS  SSSSS
    7   .FRRRRRRRRRFFFFFFFFFFFFFFFFFFFFFMMMMMMMMMMSSSSSSSSSSSSSSSSSSSSSSS  SSSSS
    6   .FRRRRRRRRREFFFFEFFFFFIFFFFIFFFFFFMMMMMMMMMSSSSSSSSSSSSSSSSSSSSSS  SSSSS
    5   .FRRRRRRRRRFFFFFFFFFFFFFFFFFFFFFFMMMMMMMMMMSSSSSSSSSSSSSSSSSSSSSS  SSSSS
    4   .FRRRRRRRRREFFFFEFFFFFIFFFFIFFFFFMMMMMMMMMMSSSSSSSSSSSSSSSSSSSSSS  SSSSS
    3   .FRRRRRRRRRFFFFFFFFFFFFFFFFFFFFFFFMMMMMMMMMMSSSSSSSSSSSSSSSSSSSSS  SSSSS
    2   .FRRRRRRRRREFFFFEFFFFFIFFFFIFFFFFFFFMMMMMMMMMSSSSSSSSSSSSSSSSSSSS  SSSSS
    1   ..................................................................  .....
        12345678901234567890123456789012345678901345678901234567890123456789012345  23456
                 10        20        30        40        50        60         146
NOTE: R-RESIDENTIAL AREA, E-CANDIDATE EXTRACTION, I- CANDIDATE INJECTION, E-OPTIMAL EXTRACTION, I-OPTIMAL INJECTION
```

Figure 6.24: The SW toe and FW tip simulated with the twentieth realization of the hydraulic conductivity after the first iteration.

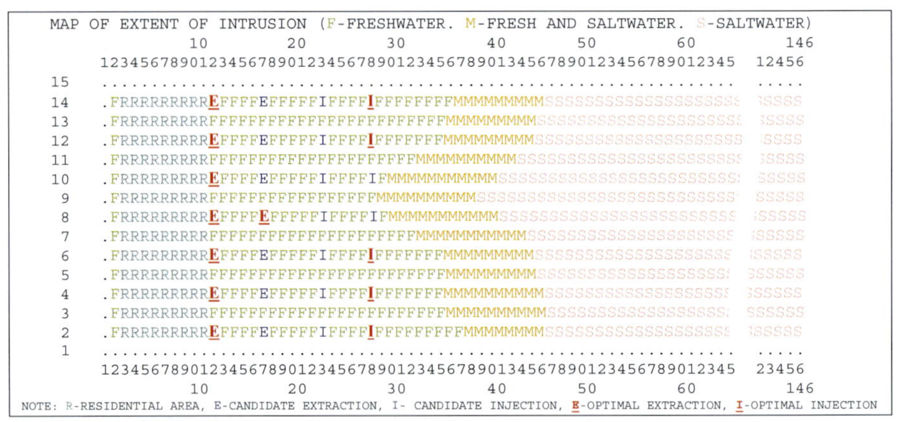

```
   MAP OF EXTENT OF INTRUSION (F-FRESHWATER. M-FRESH AND SALTWATER. S-SALTWATER)
                 10        20        30        40        50        60         146
        12345678901234567890123456789012345678901345678901234567890123456789012345  12456
   15   ..................................................................  .....
   14   .FRRRRRRRRREFFFFEFFFFFIFFFFIFFFFFFFFFMMMMMMMMMSSSSSSSSSSSSSSSSSSSS  SSSSS
   13   .FRRRRRRRRRFFFFFFFFFFFFFFFFFFFFFFFFMMMMMMMMMSSSSSSSSSSSSSSSSSSSSS  SSSSS
   12   .FRRRRRRRRREFFFFEFFFFFIFFFFIFFFFFFFMMMMMMMMMSSSSSSSSSSSSSSSSSSSSS  SSSSS
   11   .FRRRRRRRRRFFFFFFFFFFFFFFFFFFFFFMMMMMMMMMMSSSSSSSSSSSSSSSSSSSSSSS  SSSSS
   10   .FRRRRRRRRREFFFFEFFFFFIFFFFIFMMMMMMMMMMMSSSSSSSSSSSSSSSSSSSSSSSSS  SSSSS
    9   .FRRRRRRRRRFFFFFFFFFFFFFFFFFMMMMMMMMMMSSSSSSSSSSSSSSSSSSSSSSSSSSS  SSSSS
    8   .FRRRRRRRRREFFFFEFFFFFIFFFFIFMMMMMMMMMMMSSSSSSSSSSSSSSSSSSSSSSSSS  SSSSS
    7   .FRRRRRRRRRFFFFFFFFFFFFFFFFFFFFFMMMMMMMMMMMSSSSSSSSSSSSSSSSSSSSSS  SSSSS
    6   .FRRRRRRRRREFFFFEFFFFFIFFFFIFFFFFFFMMMMMMMMMSSSSSSSSSSSSSSSSSSSSS  SSSSS
    5   .FRRRRRRRRRFFFFFFFFFFFFFFFFFFFFFFFFMMMMMMMMMSSSSSSSSSSSSSSSSSSSSS  SSSSS
    4   .FRRRRRRRRREFFFFEFFFFFIFFFFIFFFFFFFMMMMMMMMMMSSSSSSSSSSSSSSSSSSSS  SSSSS
    3   .FRRRRRRRRRFFFFFFFFFFFFFFFFFFFFFFFMMMMMMMMMMMSSSSSSSSSSSSSSSSSSSS  SSSSS
    2   .FRRRRRRRRREFFFFEFFFFFIFFFFIFFFFFFFFFMMMMMMMMSSSSSSSSSSSSSSSSSSSS  SSSSS
    1   ..................................................................  .....
        12345678901234567890123456789012345678901345678901234567890123456789012345  23456
                 10        20        30        40        50        60         146
NOTE: R-RESIDENTIAL AREA, E-CANDIDATE EXTRACTION, I- CANDIDATE INJECTION, E-OPTIMAL EXTRACTION, I-OPTIMAL INJECTION
```

Figure 6.25: The SW toe and FW tip simulated with the twentieth realization of the hydraulic conductivity after the third iteration.

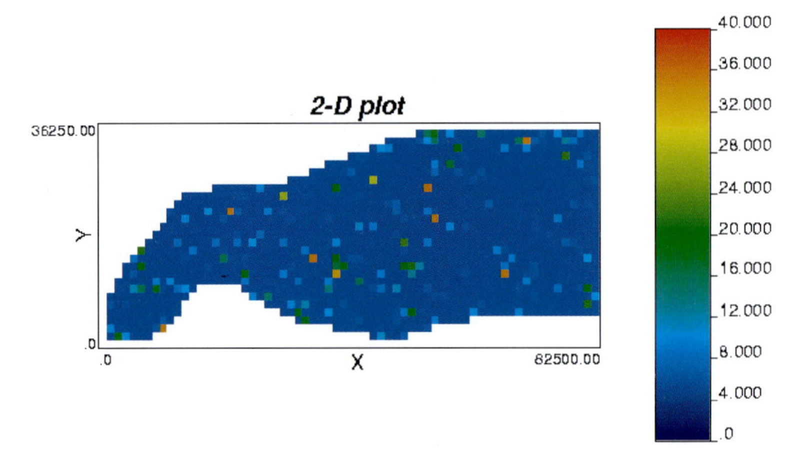

Figure 7.19. The hydraulic conductivity distribution of the 20th realization.

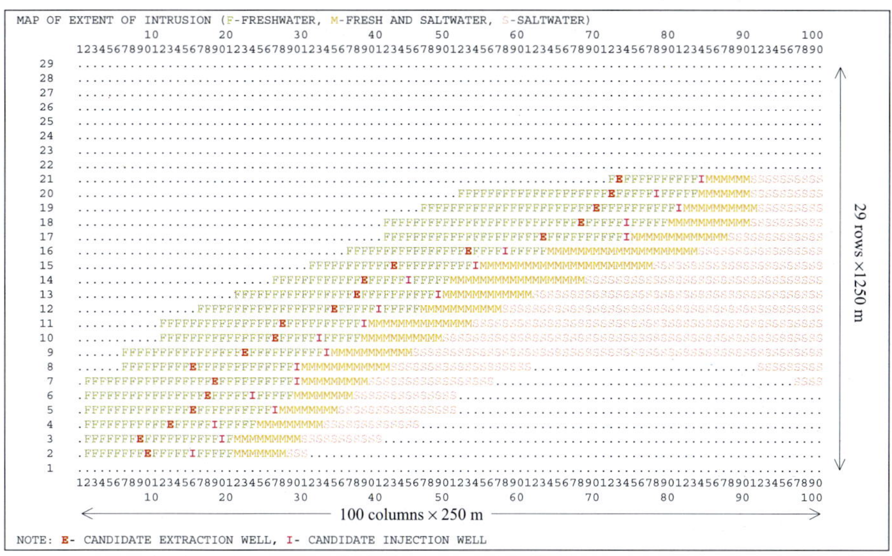

Figure 7.4: The interface toe and tip in the modeled area in the case of non-pumping condition (k = 5 m/d).

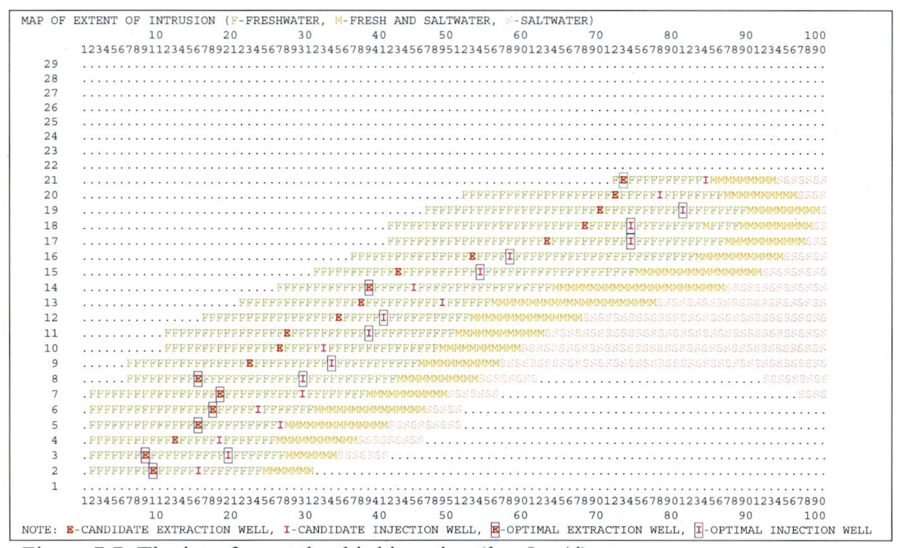

Figure 7.7: The interface at the third iteration (k = 5 m/d).

```
MAP OF EXTENT OF INTRUSION (F-FRESHWATER, M-FRESH AND SALTWATER, S-SALTWATER)
          10        20        30        40        50        60        70        80        90       100
29  ....................................................................................................
28  ....................................................................................................
27  ....................................................................................................
26  ....................................................................................................
25  ....................................................................................................
24  ....................................................................................................
23  ....................................................................................................
22  ....................................................................................................
21  .......................................................................FEFFFFFFFFFFIMMMMMMMMMSSSSSSS
20  ...................................................FFFFFFFFFFFFFFFFFFFFEFFFFFIFFFFFFFMMMMMMMMMMMSSSS
19  ..............................................FFFFFFFFFFFFFFFFFFFFFFFEFFFFFFFFFFIFFFFFFFFMMMMMMMMMMS
18  .........................................FFFFFFFFFFFFFFFFFFFFFFFFFFEFFFFFIFFFFFFFFFMFFFFMMMMMMMMMMSS
17  .........................................FFFFFFFFFFFFFFFFFFFFFEFFFFFFFFFFIFFFFFFFFFFFFMMMMMMMMMMMSSS
16  ....................................FFFFFFFFFFFFFFFFEFFFFIFFFFFFFFFFFFFFFFFFFFFFFFMMMMMMMMMMMMSSSSSS
15  ...............................FFFFFFFFFFFEFFFFFFFFFFIFFFFFFFFFFFFFFFFFFFFMMMMMMMMMMMMMMMMMSSSSSSSSS
14  ..........................FFFFFFFFFFFFEFFFFFIFFFFFFFFFFFFFFFFFFFMMMMMMMMMMMMMMMMMMMMMMSSSSSSSSSSSSSS
13  .....................FFFFFFFFFFFFFFFFEFFFFFFFFFFIFFFFFFFFFFFMMMMMMMMMMMMMMMMMSSSSSSSSSSSSSSSSSSSSSSS
12  ................FFFFFFFFFFFFFFFFFFEFFFFFIFFFFFFFFFFFFFFFFMMMMMMMMMMMMMSSSSSSSSSSSSSSSSSSSSSSSSSSSSSS
11  ...........FFFFFFFFFFFFFFFFEFFFFFFFFFFIFFFFFFFFFFFFFFFMMMMMMMMMMMMSSSSSSSSSSSSSSSSSSSSSSSSSSSSSSSSSS
10  ...........FFFFFFFFFFFFFFFEFFFFFIFFFFFFFFFFFFFFFFFFFFMMMMMMMMMMMSSSSSSSSSSSSSSSSSSSSSSSSSSSSSSSSSSSSSS
 9  ......FFFFFFFFFFFFFFFFEFFFFFFFFFFIFFFFFFFFFFFFFFMMMMMMMMMMSSSSSSSSSSSSSSSSSSSSSSSSSSSSSSSSSSSSSSSSSSS
 8  ......FFFFFFFFFEFFFFFFFFFFFFFIFFFFFFFFFFFFFFFMMMMMMMMMMSSSSSS..............................SSSSSSSSS
 7  .FFFFFFFFFFFFFFFFFEFFFFFFFFFFIFFFFFFFFFFFFMMMMMMMMMMSSSS........................................SSSS
 6  .FFFFFFFFFFFFFFFFEFFFFFIFFFFFFFFFFFFFFMMMMMMMMMMMSS.................................................
 5  .FFFFFFFFFFFFFFEFFFFFFFFFFIFFFFFFFMMMMMMMMMMMSSSSSS.................................................
 4  .FFFFFFFFFFFEFFFFFIFFFFFFFFFFFMMMMMMMMMMMSSSSS......................................................
 3  .FFFFFFFEFFFFFFFFFFIFFFFFFFFMMMMMMMMMSSSS...........................................................
 2  .FFFFFFFFEFFFFFIFFFFFFFFFMMMMMM.....................................................................
 1  ....................................................................................................
    1234567891123456789212345678931234567894123456789512345678901234567890123456789012345678901234567890
             10        20        30        40        50        60        70        80        90       100
NOTE: E-CANDIDATE EXTRACTION WELL, I-CANDIDATE INJECTION WELL, E-OPTIMAL EXTRACTION WELL, I-OPTIMAL INJECTION WELL
```

Figure 7.10: The interface at the second iteration (k = 5 m/d).

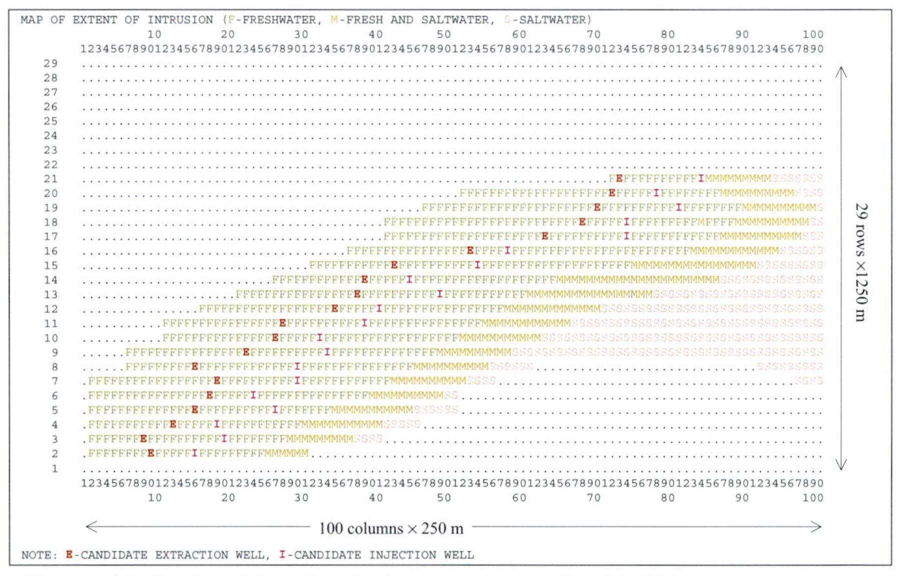

Figure 7.13: The initial interface for the second stage (k = 5 m/d).

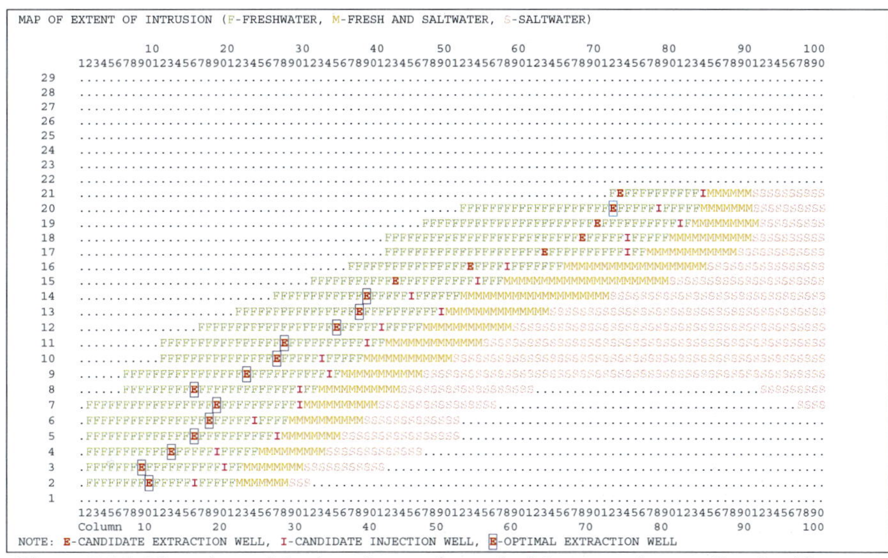

Figure 7.16: Plan view of the interface at the end of the second stage (k = 5 m/d).

MAP OF EXTENT OF INTRUSION (F-FRESHWATER, M-FRESH AND SALTWATER, S-SALTWATER)

```
Column  10        20        30        40        50        60        70        80        90       100
1234567890123456789012345678901234567890123456789012345678901234567890123456789012345678901234567890
```

NOTE: E-CANDIDATE EXTRACTION WELL, I-CANDIDATE INJECTION WELL, E-OPTIMAL EXTRACTION WELL

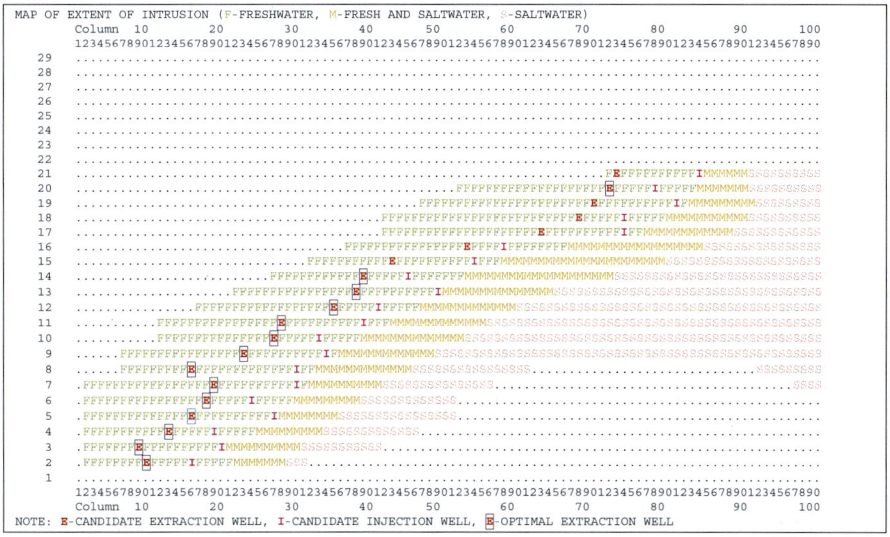

Figure 7.18: Plan view of the interface at the end of the second stage (k = 5 m/d).

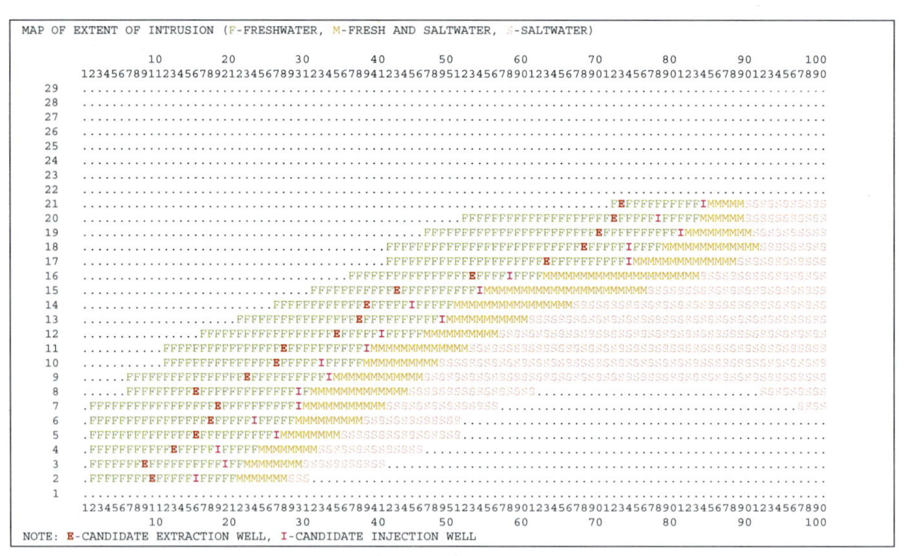

Figure 7.21. The interface at the non-pumping condition for the wet season (for the 20th realization).

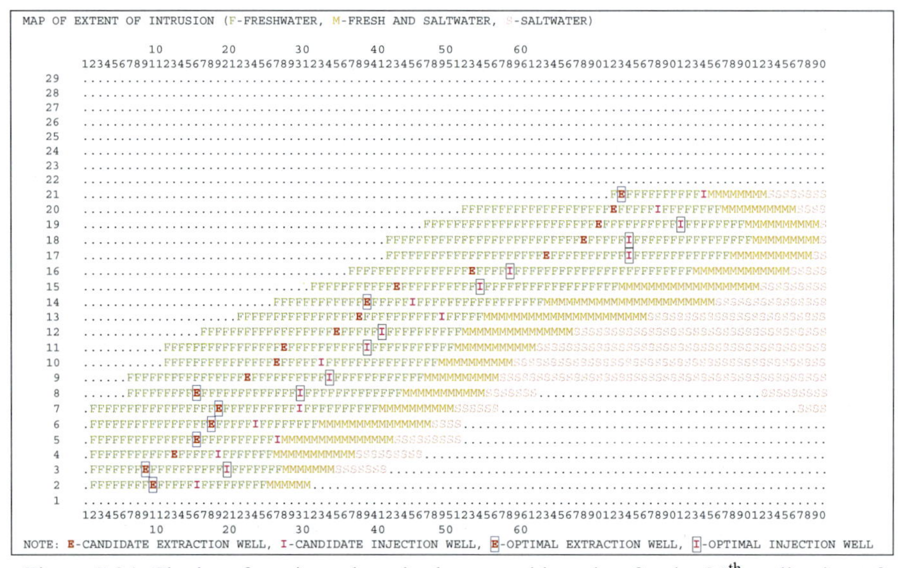

Figure 7.24: The interface tip and toe in the second iteration for the 20th realization of the hydraulic conductivity values.

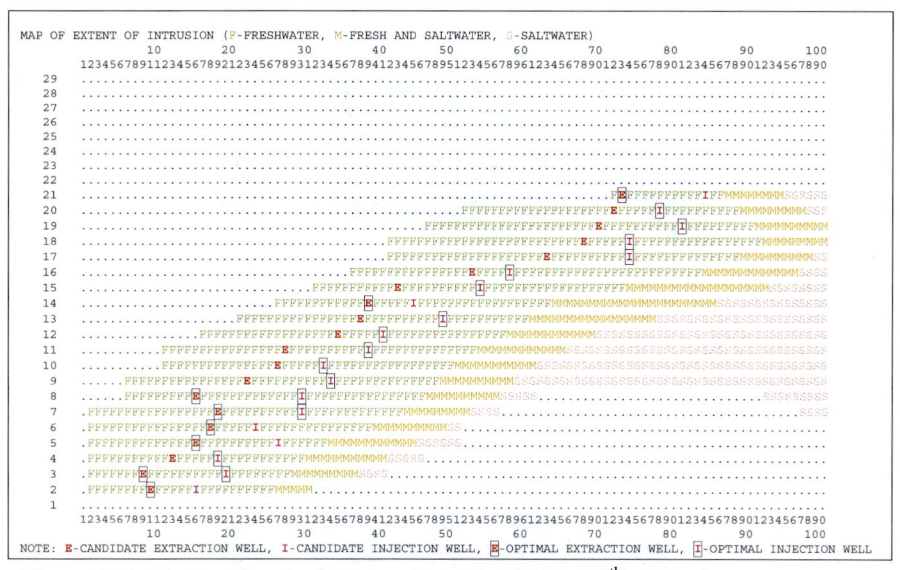

Figure 7.27: The interface in the second iteration for the 20th realization.

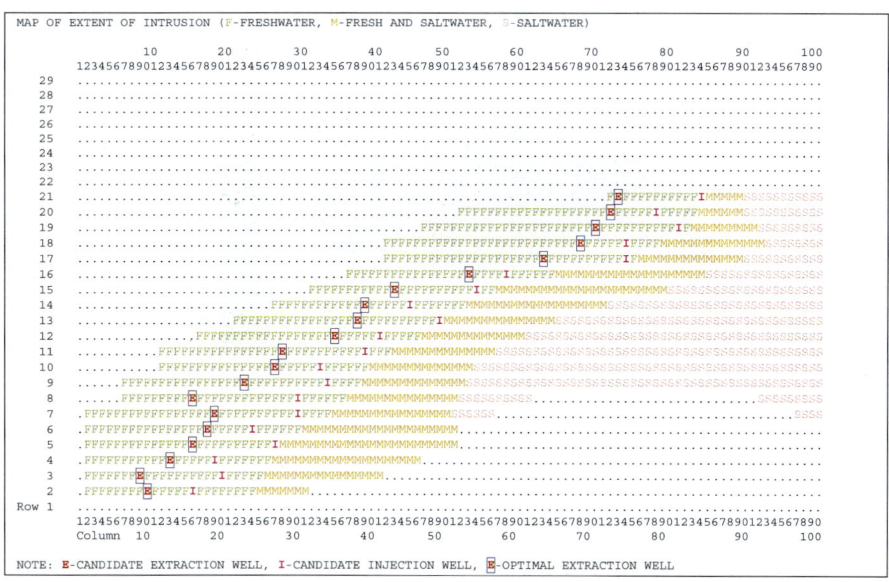

Figure 7.29: Plan view of the interface at the end of the second stage (for the 20th realization).

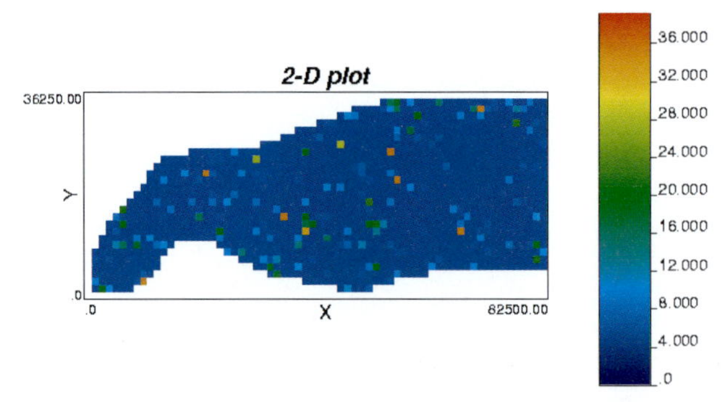

Figure 7.30: One realization of hydraulic conductivity with kriging variance = 1.

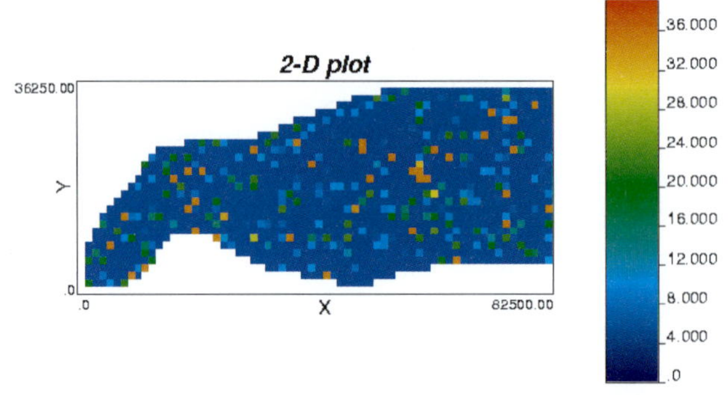

Figure 7.32: One realization of hydraulic conductivity with kriging variance = 2.

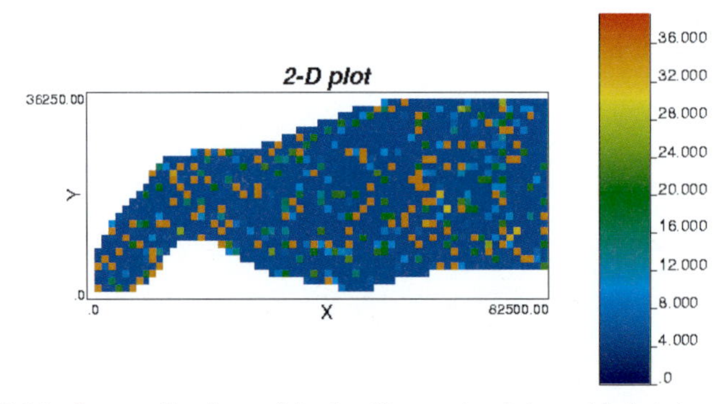

Figure 7.34: One realization of hydraulic conductivity with kriging variance = 4.

Part II

Bibliography

Bibliography

[1] Aguado, E., Sita, N., and Remson, I. (1977). "Sensitivity analysis in aquifer studies". Water Resour. Res., 13(4), 733-737.

[2] Ahlfeld, D.P. (1990). "Two-stage groundwater remediation design". J. Water Resour. Planning and Management, ASCE, 116(4), 517-529.

[3] Ahlfeld, D.P., and Heidari, M. (1994). "Applications of optimal hydraulic control to groundwater system". J. Water Resour. Planning and Management, 120(3), 350-375.

[4] Ahlfeld, D.P., and Mulligan, A.E. (2000). Optimal Management of Flow in Groundwater Systems.

[5] Ahlfeld, D.P., and Sawyer, C.S. (1990). "Well location in capture zone design using simulation and optimization techniques." Groundwater, 28(4), 507-512.

[6] Andersen, P.F., Mercer, J.W., and White, H.O. (1988). "Numerical modeling of saltwater intrusion at Hallandale, Florida". Ground Water, 26(5), 619-630.

[7] Anderson, H.R. (1978). "Hydrogeologic Reconnaissance of The Mekong Delta in South Vietnam and Cambodia". Library of Congress Cataloging in Publication Data (Geological Survey Water-Supply Paper 1608-R), USA.

[8] Anderson, M.P., and Woesner, W.W. (1992), Applied groundwater modeling - simulation of flow and advective transport. Academic Press, INC. UK.

[9] Andrecevic, R., and Kitanidis, P.K. (1990). "Optimization of the pumping schedule in aquifer remediation under uncertainty". Water Resour. Res., 26(5), 875-885.

[10] Ayers, J.F., and Vacher, H.L. (1983). "A numerical model describing unsteady flow in a fresh water lens". Water Resour. Bull., 19(5), 785-792.

[11] Barends, F.B.J., and Uffink, G.J.M. (1997). Groundwater mechanics, flow and transport. Geotechnical Laboratory, Hydraulic and Geotechnical Engineering, Faculty of Civil Engineering, Ctwa332. TU Delft, The Netherlands.

[12] Bear, J. (1979). Hydraulics of groundwater. McGraw-Hill, Inc., New York, N.Y.

[13] Ben-Tal, A., and Nemirovski, A. (1998). Convex Optimization in Engineering. Technion-Israel Institute of Technology. The lecture note consisting of 227 pages.

[14] Ben-Tal, A., and Nemirovski, A. (2001). Lectures on Modern Convex Optimization: Analysis, Algorithms; Engineering Applications. *SIAM-MPS Series on Optimization.*

[15] Bentley, L.R. (1994). "Solving and calibrating groundwater flow systems with the penalty method". Stochastic and statistical methods in Hydrology and Environmental Engineering, 2, 55-77.

[16] Bredehoeft, J.D., and Young, R.A. (1983). "Conjunctive use of groundwater and surface water for irrigated agriculture: Risk aversion". Water Resour. Res., 19(5), 1111-1121.

[17] Bruch, E.K. (1991). The Boundary Element Method for Groundwater Flow. Lecture Notes in Engineering, No. 70, Springer-Verlag Berlin, Heidelberg.

[18] Chau, T.S. (1988). "Analysis of Sustained Ground-Water Withdrawals by The Combined Simulation-Optimization Approach". Alberta Environment, Edmonton, Alberta, Can. Ground Water, 26 (4), 454-463.

[19] Cohon, J.L. (1978). Multiojective Programming and Planning. Academic Press. 333 pages.

[20] Contractor, D.N. (1983). "Numerical modeling of saltwater intrusion in the northern Guam lens". Water Resour. Bull., 19(5), 745-751.

[21] Custodio, E. and Galofre, A. (1992). Study and modeling of saltwater intrusion into aquifers. Proceedings of the saltwater intrusion meeting, 1-6 November 1992, Barcelona, Spain. CIMNE Publisher.

[22] Deutsch, Clayton.V., and André G. Journel. (1998). GSLIB Geostatistical Software Librairy and User's Guide. Second Edition. New York Oxford, Oxford University Press.

[23] Emch, P.G., and Yeh, W.W.G. (1998). "Management model for conjunctive use of coastal surface water and groundwater". J. Water Resources Planning and Management, 124(3), 129-139.

[24] Ermoliev, Yu, and Wets, R.J.B. (1988). Numerical Techniques for Stochastic Optimization. Springer, Berlin.

[25] Essaid, H.I. (1986). "A comparison of the coupled freshwater-saltwater flow and the Ghyben-Herzbergh sharp interface approaches to modeling of transient behavior in coastal aquifer systems". J. Hydrology, 86, 169-193.

[26] Essaid, H.I. (1987). Fresh water-salt water flow dynamics in coastal aquifer systems: Development and application of a multi-layered sharp interface model, Ph.D. thesis, Stanford University, California, USA.

[27] Essaid, H.I. (1990). "A multilayered sharp interface model of coupled freshwater and saltwater flow in coastal systems: Model development and applications". Water Resour. Res., 26(7), 1431-1454.

[28] Essaid, H.I. (1990). The computer model SHARP, a quasi-three dimensional finite-difference model to simulate freshwater and saltwater flow in layered coastal aquifer system: model development and applications. Rep. No. 90-4130, U.S. Geological Survey Water Resour. Investigation.

[29] Essaid, H.I. (1990a). "The computer model SHARP, a quasi-three dimensional finite-difference model to simulate freshwater and saltwater flow in layered coastal aquifer system: model development and applications". Rep. No. 90-4130, U.S. Geological Survey Water Resour. Investigation.

[30] Essaid, H.I. (1990b). "A multilayered sharp interface model of coupled freshwater and saltwater flow in coastal systems: Model development and applications". Water Resour. Res.,26(7), 1431-1454.

[31] Finney, B.A., Samsuhadi and Willis, R. (1992). "Quasi-three dimensional optimization model of Jakarta Basin". J. Water Resour. Plng. and Mgmt., ASCE, 118(1), 18-31.

[32] Gelhar,W.L. (1993). Stochastic Subsurface Hydrology, Prentice Hall, Englewood Cliffs, New Jersey.

[33] Ghaoui, L.E., and Lebret, H. (1997). "Robust solutions to Least-squares problems with uncertain data". SIAM J.Matrix Anal. Appl. Vol. 18, No 4, pp. 1035-1064, October 1997.

[34] Gharbi, A., and Peralta, R.C. (1994). "Integrated embedding optimization applied to Salt Lake valley aquifers". Water Resour. Res. 30(3), 817-832.

[35] Glover, R.E. (1959). "The pattern of fresh-water flow in a coastal aquifer". J. Geophys. Res., 64(4), 457-459.

[36] Gorelick, S.M. (1982). "A model for managing sources of groundwater pollution". Water Resour. Res., 18(4), 773-781.

[37] Gorelick, S.M. (1983). "A review of distributed parameter groundwater management modeling methods". Water Resour. Res., 19(2), 305-319.

[38] Gorelick, S.M., Voss, C.I., Gill, P.E., Murray, W., Saunders, M.A., and Wright, M.H. (1984). "Aquifer reclamation design: the use of contaminant transport simulation combined with nonlinear programming". Water Resour. Res., 20(4), 415-427.

[39] Graham, W.D., and McLaughlin, D.B. (1991). "A Stochastic model of solute transport in groundwater: application to the Borden, Ontario, tracer test". Water Resour. Res., 27(6), 1345-1359.

[40] Güler, O., and Tunçel, L. (1998). "Characterization of the barrier parameter of homogeneous convex cones". *Mathematical Programming*, 81, 55-76.

[41] Hantush, M.M.S., and Marino, M.A. (1989). "Chance-constrained model for management of a stream-aquifer system". J. Water Resour. Plann. Manage., 115(3), 259-277.

[42] Henry, H.R. (1959). "Salt intrusion into fresh-water aquifers". J. Geophys. Res., 64(11)1911-1919.

[43] Hillier, F.S. and Lieberman, G.J. (1995). Introduction to Operations Research, 6th ed., McGraw-Hill, Inc.

[44] Hobbs, B.F., Von Patterson, C., Maciejowski, M.E., and Haimes, Y.Y. (1988). "Risk analysis of aquifer contamination by brine". J. Water Resour. Plann. Manage., 114(6), 667-685.

[45] Huyakorn, P.S., Y.S. Wu, and N.S. Park, (1996). "Multiphase approach to the numerical solution of a sharp interface saltwater intrusion problem". Water Resour. Res., 32(1), 93-102.

[46] Kaunas, J.R., and Haimes, Y.Y. (1985). "Risk management of groundwater contamination in a multiobjective framework". Water Resour. Res., 21(11), 1721-1730.

[47] Konikow, L.F., and Bredehoeft, J.D.(1978). Computer model of two-dimensional solute transport and dispersion in groundwater. U.S.G.S. Techniques of Water Resources Investigations, Book 7, Chapter C2, p. 90.

[48] Kreft, A., and Zuber A. (1978). "On the physical meaning of the dispersion equation and its solutions for different initial and boundary conditions". Chemical Engineering Science, 33, 1471-1480.

[49] Lehr, J.H., and Nielsen , D.M. (1982). "Aquifer restoration and ground-water rehabilitation: A light at the end of the tunnel". Groundwater, 20(6), 650-656.

[50] Lichtenberg, E., Zilberman, D., and Bogen, K. T. (1989). "Regulating environmental health risks under uncertainty: Groundwater contamination in California". J. Environ. Econ. Manage., 17, 23-34.

[51] Liu Peimin, Liu Zhenfan and Duan Zengshan. (1994). A case study on Artificial Recharge of Groundwater into the Coastal Aquifer in Longkou, China. Artificial Recharge of Groundwater II.

[52] Maddock, T. (1974). "The operation of a stream-aquifer system under stochastic demands". Water Resour. Res., 10(1), 1-10.

[53] McDonald, M.G. and Harbaugh, A.W. (1984). A modular three-dimensional finite difference groundwater flow model, U.S.Geological Survey, Reston Virginia.

[54] Michael, Eugene D. (1971). "Use of Groundwater in Developing The Mekong Delta, Republic of Vietnam". Earth Science Research Corporation, P.O. Box 5427, Santa Monica, California 90405.

[55] Morley, C.G., et al. (1974). The Allocative Conflicts in Water-Resource Management. Agassiz center for water studies, the University of Manitoba, Canada.

[56] Ndambuki J.M. (2001). "Multi-objective groundwater quantity management: a stochastic approach". PhD thesis, Delft University of Technology, Delft.

[57] NEDECO. (1991). Mekong Delta Master Plan; (VIE/87/031). Working Paper No.2; "Groundwater Resources and Water Supply".

[58] Oude Essink, G.H.P. (1996). Impact of sea level rise on groundwater flow regimes, a sensitive analysis for the Netherlands. PhD Dissertation Thesis, volume 7, Delft studies in Integrated Water Management, Delft University of Technology. The Netherlands.

[59] Ranijthan, S., Eheart, J. W., and Garrett, J. H. (1990). Application of a neural network in groundwater remediation under conditions of uncertainty. Paper presented at International Workshop on New Uncertainty Concepts in Hydrology and Water Resources, Int. Assoc. of Hydrol. Sci., Madralin, Poland.

[60] Schotting, R.J. (1998). Mathematical Aspects of Salt transport in Porous Media. PhD dissertation thesis, CWI, Amsterdam, The Netherlands.

[61] Shamir, U., and Dagan, G. (1971). "Motion of the seawater interface in coastal aquifers: a numerical solution ". Water Resour. Res., 7(3), 644-657.

[62] Shamir, U., Bear, J., and Gamliel, A. (1984). "Optimal annual operation of a coastal aquifer". Water Resour. Res., 20(4), 435-444.

[63] Souza, W.R., and Voss, C.I. (1987). "Analysis of anisotropic coastal aquifer system using variable-density flow and solute transport simulation". J. Hydrology, 92, 17-41.

[64] Stakelbeek, A. (1999). Movement of Brackish Groundwater Near a Deep-Well Infiltration System in the Netherlands. Seawater intrusion in coastal aquifers: concepts, methods and practice / ed. by Jacob Bear...[et al.].

[65] Steuer, Ralph E. (1986). Multiple Criteria Optimization: Theory, Computation, and Application. John Wiley & Sons, Inc.

[66] Sturm, J.F. (1998-2001). " Using SEDUMI 1.02, A MATLAB TOOLBOX FOR OPTIMIZATION OVER SYMETRIC CONES" (updated for version 1.05). Department of Econometrics, Tilburg University, Tilburg The Netherlands, August 1998- October 2001.

[67] Sturm, J.F. (2002). Implementation of Interior Points Methods for Mixed Semidefinite and Second Order Cone Optimization Problems. *Optimization Methods and Software*, Vol. 17, No. 6, pp.1105-1154.

[68] Tung, Y.K. (1986). "Groundwater management by a chance constrained model". J. Water Resour. Plann. Manage., 112(1), 1-19.

[69] Uffink, G.J.M. (1990). Analysis of dispersion by the random walk method. Ph.D. dissertation thesis, Delft University of Technology, The Netherlands.

[70] Van Genuchten, M. Th., and Alves, W.J. (1982). Analytical solutions of the one-dimensional convective-dispersive solute transport equation. U.S. Department of Agriculture. Tech. Bull. No. 1661.

[71] Volker, R., and Rushton, K. (1982). "An assesment of the importance of some parameters for seawater intrusion in aquifers and a comparison of dispersive and sharp-interface modeling approaches". J. Hydrology, 56, 239-250.

[72] Wagner, B.J., and Gorelick, S.M. (1986). "A statistical methodology for estimating transport parameters: Theory and applications to one-dimensional advective-dispersive systems". Water Resour. Res., 22(8), 303-315.

[73] Wagner, B.J., and Gorelick, S.M. (1987). "Optimal groundwater quality management under parameter uncertainty". Water Resour. Res., 23(7), 1162-1174.

[74] Wagner, B.J., and Gorelick, S.M. (1989). "Reliable aquifer remediation in the presence of spatially variable hydraulic conductivity: From data to design". Water Resour. Res., 25(10), 2211-2225.

[75] Wagner, J.M. and Shamir, U. and Nemati, H.R. (1992). "Groundwater quality management under uncertainty: Stochastic programming approaches and the value of information". Water Resour. Res., 28(5), 1233-1246.

[76] Wagner, B.J. (1988). Optimal groundwater quality management under uncertainty. PhD. Thesis, Dept. of Appl. Earth Sci., Stanford Univ., Stanford, California.

[77] Willis, R. (1979). "A planning model for the management of the groundwater quality". Water Resour. Res., 15(6), 1305-1312.

[78] Willis, R., and Finney, B.A. (1985). "Optimal control of nonlinear groundwater Hydraulics: Theoretical development and numerical experiments". Water Resour. Res., 21(10), 1476-1482.

[79] Willis, R., and Finney, B.A. (1988). "Planning model for optimal control of saltwater intrusion". J. Water Resour. Plng. and Mgmt., ASCE, 110(3), 333-347.

[80] Willis, R., and Yeh, W.W-G. (1987). Groundwater system planning and management. Prentice-Hall, Inc., Englewood Cliffs, N.J. USA.

[81] Wilson, J., and Sa da Costa, A., (1982). "Finite element simulation of a saltwater/freshwater interface with indirect toe tracking". Water Resour. Res., 18(4), 1069-1080.

[82] Ye, Y., Tood, M.J., and Mizuno, S. (1994). "An$\left\{O\left(\sqrt{n}L\right)\right\}$-iteration homogeneous and self-dual linear programming algorithm". Mathematics of Operational Research, **19**, 53-67.

[83] Yoon, Jae-Heung and Shoemaker, C.A., Fellow, ASCE. (1999). "Comparison of Optimization Methods for Ground-Water Bioremediation". Journal of Water Resources Planning and Management, Vol. 125, No. 1. ©ASCE, ISSN 0733-9496/99/0001-0054-0063. Paper No. 14732, January/February, 1999.

Part III

Summary, Samenvatting and Curriculum vitae

Summary

"Multi-Objective Management of Saltwater Intrusion in Groundwater: Optimization under Uncertainty"

by Tran Minh Thuan

Coastal aquifers are very vulnerable to seawater intrusion through, for example, the overdraft of groundwater exploitation or insufficient recharge from upstream. Problems of salt-intrusion into groundwater have become a considerable concern in many countries with coastal areas. There have been a number of studies that have tried to simulate groundwater flow system in regions under threat of saltwater intrusion into coastal aquifers. These aquifer systems are characterized by either a single layer (unconfined) or multiple layers with varying hydraulic properties. These are necessary parameter inputs of such simulation models. In order to control saltwater intrusion, planning and management models have been reported in the literature under varying forms of saltwater intrusion management models. These have been based on combinations of the simulation models and optimization programs. They address optimal groundwater pumping and recharge schedules with or without surface water supply for conjunctive use. Saltwater intrusion management problems are necessary to be multi-objective. In this work, the multi-objective management schemes that are based on the minimization of the operational costs and the saltwater intrusion length as their objectives are proposed for the first time. In the literature, these saltwater intrusion management models are mostly based on deterministic aquifer parameters, e.g. the transmissivity is assumed to be precisely known. Moreover, in reality the aquifer parameters cannot be described as constants because they vary spatially. Therefore, the output of deterministic simulation models gives only a first impression of the saltwater intrusion problem. In this thesis also the uncertainty of the input parameters is taken into account in the computations. Hence, the stochastic optimization approach will be introduced to the multi-objective management of saltwater intrusion in groundwater. Moreover, the relationships of the saltwater intrusion lengths (tip and toe) with respect to stresses have a non-linear nature. Thus, their response matrices, which are the coefficients of the objective functions and constraints, are not fixed. This introduces even more complexity to the linear programming in the sense of determining the exact coefficients for the optimization problems. In this thesis, multi-objective management models are developed for a single-layered confined aquifer system. The mean value of the transmissivities is used in the deterministic management problem and realizations of random values of the hydraulic conductivity within a given range are used for the stochastic management problem. In both problems, the Second-Order Cone Optimization programming (SOCO) is applied for solving the single and multi-objective problems.

The following issues are the objectives of this thesis:

Objectives

Characterization of the responses of saltwater intrusion lengths with respect to stresses and transmissivities

Based on the SHARP computer code, the numerical experiments are performed to find the numerical relationship between the saltwater intrusion length and the boundary flow, the extraction and the injection rates in a confined aquifer where homogeneity of the hydraulic conductivity is assumed. In addition, the modeller needs to observe the response of the salt/freshwater sharp interface due to the variation of the hydraulic conductivity values, which are altered for the simulation model in the input files. This step is essential in model applications for management through the understanding of the salt/fresh water interface movement. Generally, through this sensitivity study, the non-linear response of the saltwater intrusion length with respect to stresses (extraction, injection rates) is verified and the distinct changes of the salt/fresh sharp interface with respect to the hydraulic conductivity variation are determined. These changes will cause errors in predicting the location of the interface to the management decisions if it is not realized how uncertain the distribution of the values of hydraulic conductivity is. Therefore, this results in a requirement to estimate the uncertainty of the parameters. By characterizing and verifying these response relationships, the response matrices will be computed through this simulation model under a set of input files comprised of these driving parameters.

Introduction of the application of the second-order cone optimization (SOCO) programming technique and SeDuMi (an add-on for MATLAB) into the optimal management of saltwater intrusion in groundwater

The second-order cone optimization (SOCO) (or quadratic cone) programming technique has been developed to solve a class of convex optimization problems. Nowadays this programming, together with the interior point method, is a promising method for solving large-scale optimization problems. This can be conveniently developed for the saltwater intrusion management problems, especially in cases where the coefficients of the objectives and constraints are in uncertainty fields. Besides that, SeDuMi (an add-on for MATLAB, which is an optimization program package developed for linear, SOCO and semi definite programming,) enables its incorporation in a computer program for saltwater intrusion management problems.

Development of a multi-objective management of saltwater intrusion in groundwater with deterministic and stochastic approaches as an add-on program for MATLAB

With the management of an aquifer system in coastal areas where the salt/fresh interface appears near the capture zone, it is necessary to include the objective for minimizing the saltwater intrusion length during the operation of the extraction and injection wells. Besides that the other objective for minimizing the operational costs is also included. Those objectives conflict with each other in the sense of injecting the surface water in order to control the saltwater intrusion while minimizing the operational costs for both extraction and injection. The cost of extraction of fresh groundwater for minimal drinking demand is always constant in the problem. The saltwater intrusion increment upper bound constraint will play a role similar to the adjustment options for the fresh groundwater storage purposes. The management program is built by creating the linkages between the SHARP simulation model and SeDuMi optimizer through the response matrices under the MATLAB environment.

In the sense of determining the distributed parameters (e.g. the hydraulic conductivity) for the simulation model, the management problems will be established for the two cases. These are, firstly, the deterministic management program using the (assumedly) precisely known hydraulic conductivity and, secondly, the stochastic management program based on the uncertainty of the hydraulic conductivity. The multi-objectives of these two management programs are solved by the SOCO program with the methods of the minimum distance from the ideal solution and the prior assessments of weights. For such problems the non-inferior sets are graphically shown by the programs to illustrate the range of equal choice for the optimal values under different values of α in the L_α-metric problems or the weights w in the weighted problems.

Application of the programs to hypothetical and real world problems for the saltwater intrusion management

The multi-objective management program for saltwater intrusion in groundwater is firstly applied to a hypothetical case in which the geometry of the modelled area is assumed to be symmetric. Either the mean value of the hydraulic conductivity is given (the deterministic case) or realizations of the hydraulic conductivity are generated (the uncertainty case) for the input data of the simulation model. The candidate well locations, being the decision variables, are arranged in a symmetric way so that the results of the hypothetical problem in the deterministic case will help to check the validity of the programs because the optimal solution obtained should be symmetric. The results in Chapter 6 will satisfy these expectations.

The real world problem is addressed in one particular study area, selected from the coastal areas in the Mekong Delta in Viet Nam. The area is intruded by saltwater with the current interface position located near the pumping wells. For this particular area the available data are very scarce and only the averaged values of aquifer properties are given. Unlike the modelled area in the hypothetical problem, in the real world problem the modelled area has an asymmetric shape. This management scheme for the prevention of saltwater intrusion by artificial injection, as proposed here, is new and can be applied in areas where there is a potential risk of saltwater intrusion. Therefore, the saltwater intrusion management problem presented in this work will give more insight in the required measures for the future rather than the current management scheme of the study area.

Since the study area is subject to the tropical monsoon weather, there are two distinct seasons, the wet and the dry seasons. Under these circumstances a scheme of management is proposed – in this sense a so-called seasonal planning for the saltwater intrusion management problem. This consists of two managerial stages during one year– the first stage during the wet season and the second stage during the dry season. The program will run for the first stage and its solution that depends on the L-metric or weight values of the two objectives will be attained. The salt/fresh interface that is simulated using the model with the optimal solution obtained in the first stage will be the initial interface for the second stage of the management problem. The results in Chapter 7 show that optimal solutions of both L_2-metric and weighted problems can be obtained for the first stage management scheme. These optimal values will be found reasonably different between the deterministic and uncertainty cases. For the second stage, however, because the chosen objectives are not conflicting with each other for this stage all non-inferior sets have alternative optima. Therefore the optimal solutions found can be only local optimal solutions that approach the utopia points.

Samenvatting

"Op meervoudige doelstellingen gebaseerd beheer van zoutwaterintrusie in grondwater: optimalisatie en onzekerheid"

door Tran Minh Thuan

Kustaquifers zijn zeer kwetsbaar voor intrusie van zeewater. Problemen van zoutwaterintrusie in grondwater zijn een bijzonder aandachtspunt geworden in veel landen met kustgebieden. In een aantal studies heeft men geprobeerd grondwaterstroming te simuleren in gebieden waar deze bedreiging aanwezig is. Deze aquifer systemen worden gekenmerkt door hetzij een enkele aquifer (freatisch) of door een meer-lagen systeem met verschillende hydraulische eigenschappen. Dit zijn noodzakelijke invoerparameters voor zulke simulatie modellen. In de literatuur zijn ontwerp- en beheersmodellen te vinden om intrusie van zout water te voorkomen. Deze zijn gebaseerd op combinaties van simulatie en optimalisatie programma's. Zij richten zich op het optimaliseren van het oppompen en infiltreren van grondwater met of zonder gebruik vanoppervlaktewater. Het beheer van zoutwaterindringing levert noodzakelijkerwijs problemen met tegenstrijdige doelstellingen op. In dit proefschrift worden voor de eerste keer de beheersschema's voorgesteld gebaseerd op meervoudige doelen, namelijk de minimalisatie van de operationele kosten en de minimalisatie van de zoutwaterintrusielengte. In de literatuur worden deze beheersmodellen om zoutwaterintrusie tegen te gaan meestal gebaseerd op deterministische waarden voor de aquifer parameters. Bijvoorbeeld wordt aangenomen dat de transmissiviteit precies bekend is. Daarom geven de resultaten van een deterministische simulatie slechts een eerste indruk van het probleem van zoutwaterintrusie. In dit proefschrift wordt er wel rekening gehouden met de onzekerheid van de invoerparameters. Vandaar dat hier de stochastische optimalisatie-aanpak wordt ge ntroduceerd bij deze op meervoudige doelstellingen gebaseerd beheer van zoutwaterintrusie in grondwater. Bovendien hebben de relaties tussen de zoutwaterintrusielengten (bovenin ("tip") en onderin de aquifer ("toe")) een niet-lineair karakter. Zo liggen de responsie matrices die de coëfficiënten vormen voor de doelfuncties en bijbehorende beperkingen niet vast. Dit introduceert een extra complicatie bij het lineair programmeerwerk om de exacte coëfficiënten te vinden bij de optimalisatie problemen. De gemiddelde waarde van de transmissiviteiten wordt gebruikt bij het deterministische beheersprobleem en realisaties van random waarden van de hydraulische conductiviteit binnen een gegeven bereik worden gebruikt voor het stochastische beheersprobleem. Voor beide problemen wordt de "Second-Order Cone Optimization" (SOCO) programmeertechniek toegepast zowel voor de enkele als voor de meervoudige doelstelling.

De volgende onderwerpen worden in dit proefschrift behandeld:

Doelen

Karakterisering van de responsies van de zoutwaterintrusielengten met betrekking tot drukken en transmissiviteiten

Numerieke experimenten zijn uitgevoerd gebaseerd op de SHARP computer code om de numerieke relatie te vinden tussen de zoutwaterintrusielengte en de stroming over de rand van het gebied, de onttrekking en het injectiedebiet in een afgesloten aquifer waar homogeniteit van de hydraulische conductiviteit wordt aangenomen. Bovendien dient de modelleur de responsie vast te stellen van het zoet-zoutscheidingsvlak ten gevolge van de variatie van de hydraulische conductiviteitswaarden. In het algemeen wordt door deze gevoeligheidsstudie de niet-lineaire responsie van de zoutwaterintrusielengte met betrekking tot drukken (extractie, injectiedebieten) bevestigd en worden er duidelijke veranderingen van het zoet-zoutscheidingsvlak met betrekking tot de hydraulische conductiviteitsvariatie vastgesteld. Deze veranderingen zullen fouten veroorzaken bij het voorspellen van de ligging van het scheidingsvlak bij beheersbeslissingen indien men zich niet realiseert hoe onzeker de hydraulische conductiviteitsverdeling is. Dit betekent dat men een schatting dient te hebben van deze onzekerheid. Door deze responsie relaties te karakteriseren en te verifiëren worden de responsie matrices berekend door het simulatiemodel (SHARP) met als invoer een verzameling van invoerbestanden bestaande uit informatie over deze sturende parameters.

Introductie van de toepassing van de "Second-Order Cone Optimization" (SOCO) programmeertechniek en SeDuMi (een toepassingsprogramma voor MATLAB) binnen het optimaal beheer van zoutwaterintrusie in grondwater

De "Second-Order Cone Optimization" (SOCO) (of kwadratische kegel) programmeertechniek is ontwikkeld om een klasse van convexe optimalisatie problemen op te lossen. Tegenwoordig is deze programmeertechniek te samen met de "interior point method" een veelbelovende methode om grootschalige optimalisatie problemen op te lossen. Dit kan op een geschikte manier worden ontwikkeld bij de beheersproblemen voor de zoutwaterintrusie en zeker in die gevallen waar de coëfficiënten voor de doelstellingen en de beperkingen in een onzekerheidsinterval liggen. Bovendien is het mogelijk om SeDuMi (een optimalisatie programma voor lineaire, SOCO en "semi-definite" problemen als toepassingsprogramma binnen de programmeeromgeving MATLAB) te koppelen met SHARP om zo tot een beheersprogramma te komen voor zoutwaterintrusie.

Ontwikkeling van een op meervoudige doelstellingen gebaseerd beheer van zoutwaterintrusie in grondwater met deterministische en stochastische benaderingen als een toepassingsprogramma binnen MATLAB

Bij het beheer van een aquifer systeem in kustgebieden waar het zoet-zoutscheidingsvlak vlakbij het intrekgebied van een winning ligt is het noodzakelijk als doelstelling mee te nemen de minimalisatie van de zoutwaterintrusielengte gedurende de werking van de extractie- en injectieputten. Bovendien wordt het minimaliseren van de operationele kosten meegenomen. Deze doelstellingen zijn strijdig met elkaar in de zin dat oppervlaktewater wordt geïnjecteerd om zoutwaterintrusie te voorkomen terwijl men ook de operationele kosten van de extractie- en injectieputten wilt minimaliseren. De kosten van de extractie van zoet grondwater om te voldoen aan de minimale drinkwatervraag zijn altijd constant verondersteld. De bovengrens aan de toename van de zoutwaterintrusielengte zal een rol spelen vergelijkbaar met de aanpassingsmogelijkheden voor het opslaan van zoet grondwater. Het beheersprogramma wordt samengesteld door het SHARP simulatieprogramma te koppelen met het SeDuMi optimimalisatie programma door de uitwisseling van responsie matrices binnen de MATLAB programmeeromgeving. Om de ruimtelijk toegekende parameters (bijvoorbeeld de hydraulische conductiviteit) voor het simulatieprogramma te bepalen, zal het beheersprobleem worden opgezet voor twee gevallen. Dat is allereerst het deterministische beheersprogramma dat gebruik maakt van de aangenomen precies bekende hydraulische conductiviteit en ten tweede is dat het stochastische beheersprogramma gebaseerd op de onzekerheidsvelden van de random waarden voor de hydraulische conductiviteit. De meervoudige doelstellingen van deze twee beheersprogramma's worden opgelost met het SOCO programma met daarin verwerkt de minimale afstand tot de ideale oplossing en de van te voren vastgelegde schatting voor de onderlinge gewichtsfactoren. Voor zulke soorten problemen worden de niet-inferieure verzamelingen graphisch getoond om het bereik van gelijkwaardige keuzen te laten zien voor de optimale waarden voor de verschillende α-waarden bij de L_α-norm aanpak of de gewichten w bij de aanpak met weging van de tegenstrijdige doelstellingen.

Toepassing van de programma's op hypothetische en realistische problemen voor het beheer van het zoutwaterintrusieprobleem

Het op meervoudige keuzen gebaseerde beheersprogramma voor zoutwaterintrusie in grondwater wordt eerst toegepast op een hypothetisch geval waarin de geometrie van het te modelleren gebied symmetrisch wordt verondersteld. Hetzij de gemiddelde waarde van de hydraulische conductiviteit wordt gegeven (het deterministische geval) hetzij realisaties van de hydraulische conductiviteit (het stochastische geval) worden gegenereerd als invoer gegevens voor het simulatie model. De beoogde putlocaties die optreden als beslissingsvariabelen worden op een symmetrische wijze gesitueerd opdat de resultaten van het hypothetische probleem kunnen meehelpen om de correcte werking van het programma te controleren omdat dan namelijk ook de oplossing symmetrisch dient te zijn. De resultaten in Hoofdstuk 6 voldoen aan deze verwachtingen.

Het realistische probleem is toegesneden op een studie gebied gekozen uit het kustgebied van de Mekong Delta in Viet Nam. Het gebied wordt bedreigd door binnendringend zout water terwijl het huidige zoet-zoutscheidingsvlak dichtbij de winninglocaties ligt. De beschikbare gegevens voor dit gebied zijn erg spaarzaam en enkel gemiddelde waarden voor de aquifer eigenschappen zijn bekend. In tegenstelling tot het hypothetische probleem heeft het gemodelleerde gebied nu geen symmetrische vorm. Het hier voorgestelde beheersschema is nieuw en kan toegepast worden in gebieden waar een potentiëel gevaar bestaat voor zoutwaterintrusie. Daarom zal het hier beschreven beheersschema om zoutwaterintrusie tegen te gaan meer inzicht geven in vereiste maatregelen in de toekomst dan het huidige beheersplan voor het studiegebied.

Omdat het studiegebied onderworpen is aan het tropische moesson-klimaat zijn er twee verschillende seizoenen, het natte en het droge. Onder deze omstandigheden wordt een beheersplan voorgesteld dat rekening houdt met het seizoensgebonden karakter van de zoutwaterintrusie. Dat behelst twee beheersperioden gedurende een jaar, de eerste gedurende het natte seizoen, de tweede gedurende het droge. Het programma draait voor de eerste periode en de oplossing, die afhangt van de L-norm of van de onderlinge weging van de twee doelstellingen, wordt gevonden. Het zoet-zoutscheidingsvlak dat verkregen wordt met het model als de optimale oplossing zal de beginsituatie vormen voor het beheersprobleem gedurende de tweede periode. De resultaten in Hoofdstuk 7 laten zien dat optimale oplossingen zowel voor de L_2-norm als voor de gewogen problemen verkregen kunnen worden voor de eerste fase van het beheersprobleem. Deze optimale waarden voor het deterministische en het stochastische geval zijn redelijk verschillend. Voor de tweede fase echter hebben alle niet-inferieure verzamelingen alternatieve optima omdat de gekozen doelstellingen niet met elkaar strijdig zijn. Daarom kunnen de gevonden optimale oplossingen slechts lokale optimale oplossingen zijn die de utopia punten benaderen.

Curriculum vitae

The author of this thesis, Tran Minh Thuan, was born on 26th October 1957 in Can Tho province, Viet Nam. His schooling began in 1962 and had experienced for thirteen years through the primary school in Ca Mau and the vocational secondary school in Can Tho before he followed the four-year study in Can Tho University. With the B.Sc. degree in Civil Engineering he was recruited to be a teaching staff of the faculty of Water Management in Can Tho University. After two-year working, he was called to join the army in the logistics from 1982 to 1984 as the youth duty. Completing the duty he went back to Can Tho University for teaching again. In 1987 he passed an entrance examination and was granted a fellowship to study M.Sc. course in Wageningen University, the Netherlands. Two year later he was graduated with a M.Sc. degree in Soil Science and Water Management and returned to his country for continuing the teaching task. From 1989 to 1996, besides teaching, he performed a number of projects for "the poverty alleviation program" funded by the NGO (Fado, Oxfam America and Novib). He also wrote some computer programs for consultant companies in the Mekong Delta region for designing civil engineering constructions. In 1997 he started his Ph.D. research in a sandwich program project "Development of Curricula in Civil and Mechanical Engineering", Can Tho University, paid by the MHO-program under supervision of Nuffic and coordinated by CICAT, Delft.